THE CINEMATIC CITY

•

The role played by the city is central to a wide variety of films. In everyday experience, cities frequently seem to possess a cinematic quality. Yet the relationship between city and cinema has been neglected in both film and urban studies.

The contributors to this multidisciplinary volume have sought to remedy this situation by offering a detailed and wide-ranging look at the 'cinematic city'. Covering a diverse selection of films, genres, cities, and historical periods, the innovative and thought-provoking essays gathered together here offer numerous insights to those interested in cities, film, and their conjunction.

By introducing some of the major dimensions of urban theory and film theory at the outset, the collection proceeds towards a detailed understanding of the cinematic forms most significantly related to the city – including early cinema, documentary film, *film noir*, 'New Wave' and postmodern cinema. The volume provides a wealth of empirical detail whilst drawing on the theoretical insights of Benjamin, Baudrillard, Foucault, Deleuze, Lacan, Lefebvre and contemporary feminism.

The Cinematic City is the first major collection of its kind. This unique collection of specially commissioned pieces should change the way in which we think about both the city and the cinema. The essays included here permit one to think afresh about their connection, and about the particular qualities this link affords to both.

David B. Clarke is Lecturer in Geography at the University of Leeds.

THE CINEMATIC CITY

•

Edited by

David B. Clarke

London and New York

First published 1997
by Routledge
11 New Fetter Lane, London EC4P 4EE

Simultaneously published in the USA and Canada
by Routledge
29 West 35th Street, New York, NY 10001

Typeset in Sabon by
Florencetype Limited, Stoodleigh, Devon
Printed and bound in Great Britain by
Butler & Tanner Ltd, Frome and London

British Library Cataloguing in Publication Data
A catalogue record for this book is available from the British Library

Library of Congress Cataloguing in Publication Data
The cinematic city / edited by David B. Clarke.
p. cm.
Includes bibliographical references and index.
1. Cities and towns in motion pictures. 2. City and town life in motion pictures.
I. Clarke, David B.
PN1995.9.C513C46 1997
791.43'621732—dc20 96–31750

ISBN 0–415–12745–9 (hbk)
0–415–12746–7 (pbk)

CONTENTS

•

PLATES

●

CONTRIBUTORS

●

Giuliana Bruno is Associate Professor in the Department of Visual and Environmental Studies at Harvard University. She has published widely on film and cultural theory and is the author of *Streetwalking on a Ruined Map: Cultural theory and the City Films of Elvira Notari* (Princeton University Press, 1993).

Iain Chambers teaches at the Istituto Universitario Orientale, Naples. His most recent book is *Migrancy, Culture, Identity* (Routledge, 1994), and with Linda Curti has edited *The Post-Colonial Question: Common Skies, Divided Horizons* (Routledge, 1996).

David B. Clarke is Lecturer in Human Geography at the University of Leeds and an ESRC Research Fellow. He has published on a variety of geographical topics and is currently working on a book on the geographies of the consumer society, *Commodity, Sign and Space* (Blackwell, forthcoming).

Marcus A. Doel is Lecturer in the Department of Geography at Loughborough University. He has published extensively on social–theoretical issues within geography and is currently working on a book, *Geography/Deconstruction/Postmodernism* (Guilford, forthcoming).

Antony Easthope is Professor in English and Cultural Studies at the Manchester Metropolitan University. He is the author of numerous articles and books including *British Poststructuralism* (Routledge, 1988) and *Literary into Cultural Studies* (Routledge, 1993), and has edited *Contemporary Film Theory* (Longman, 1993).

John R. Gold is Professor of Geography in the School of Social Sciences, Oxford Brookes University. He has published widely on the future city, planning and film. His work includes a bibliography on *The City in Film* (Vance Bibliographies, 1984) and a co-edited collection on *Geography, the Media and Popular Culture* (Croom Helm, 1985).

James Hay is based in the Department of Speech Communication at the College of Liberal Arts and Science, University of Illinois at Urbana–Champaign. He has published widely on film, including a book on *Popular Film Culture in Fascist Italy* (Indiana University Press, 1987).

Frank Krutnik is Senior Lecturer in Film and Television at the Roehampton Institute, London. He has written books on *film noir* and, with Steve Neale, on film and television comedy. He is currently working on a study of the career of Jerry Lewis, to be published by the Smithsonian Institution Press.

Rob Lapsley is a part-time lecturer based in Manchester. He has published a number of articles on film and is co-author, with Mike Westlake, of *Film Theory: An Introduction* (Manchester University Press, 1988).

Colin McArthur, formerly head of the distribution division of the British Film Institute, is now a freelance writer and lecturer. He has written extensively on Hollywood cinema, British television and Scottish culture. His publications include *Underworld USA* (Secker and Warburg/BFI, 1972), *Television and History* (BFI, 1978) and *Scotch Reels* (BFI, 1982). He has held visiting professorships at Queen Margaret College, Edinburgh and Glasgow Caledonian University and currently teaches at the Royal College of Art.

Elisabeth Mahoney is a lecturer in the Department of English at the University of Aberdeen. She is currently working on a book-length study of city-texts in film, photography and literature, with particular reference to the representation of sexual difference and identity in urban space.

Will Straw is Associate Professor within the Graduate Program in Communication at McGill University and Director of the Centre for Research on Canadian Cultural Industries and Institutions. He is a member of the editorial boards of *Screen*, *Cultural Studies*, *Social Semiotics*, and *Perfect Beat*, and co-editor of a new journal, *Topia: Cultural Studies in Canada*.

Stephen V. Ward is Professor of Planning History at Oxford Brookes University. His works includes numerous articles on planning, an edited collection on *The Garden City* (Spon, 1992) and *Planning and Urban Change* (Paul Chapman, 1994).

INTRODUCTION
previewing the cinematic city
•

David B. Clarke

Where is the cinema? It is all around you outside, all over the city, that marvellous,
continuous performance of films and scenarios.
BAUDRILLARD, *1988: 56*

THE CINEMATIC CITY: TAKE 1

Looking back to the European city from the heart of *America*, Baudrillard (1988:
56) writes of the 'feeling you get when you step out of an Italian or a Dutch
gallery into a city that seems the very reflection of the paintings you have just
seen, as if the city had come out of the paintings and not the other way around.'
In precisely the same way, he notes, 'The American city seems to have stepped
right out of the movies' (*ibid.*). He adds, crucially: 'To grasp its secret, you should
not, then, begin with the city and move inwards towards the screen; you should
begin with the screen and move outwards towards the city' (*ibid.*); a conceptualization of the cityscape as *screenscape*. Yet despite the immediately perceptible
cinematic qualities that cities frequently seem to possess, and despite the uncredited
role played by the city in so many films, relatively little theoretical attention has
been directed towards understanding the relationship between urban and cinematic
space (cf. Aitken and Zonn, 1994; Clarke, 1994). Indeed, whilst the histories of
film and the city are imbricated to such an extent that it is unthinkable that the
cinema could have developed without the city, and whilst the city has been unmis-
takably shaped by the cinematic form, neither film nor urban studies has paid the
warranted attention to their connection.

The city has certainly been understated in film theory. So central is the city
to film that, paradoxically, the widespread *implicit* acceptance of its importance
has mitigated against an explicit consideration of its actual significance. Indeed,
those film theorists who *have* sought to place the city in the foreground have
been widely regarded as making an innovative argument (Bruno, 1993; Friedberg,

1993). Needless to say, such a reception has to some extent worked against the broader consideration of such issues. Similarly, whilst urban and art historians have contributed to the understanding of visual representations of the city in painting (Arscott and Pollock, with Wolff, 1988; Clark, 1984; Daniels, 1986; Olsen, 1983; Pollock, 1988; Shapiro, 1984) and photography (Batchen, 1991; Tagg, 1988, 1990), relatively little writing on the city has looked in detail at the moving pictures of cinema (though see Sutcliffe, 1984; Gold 1984). Until recently, most urban theorists have displayed an almost wilful disregard for such a supposedly tangential topic as film. And, once again, those who have sought to highlight its significance have generally been regarded as pursuing a 'specialist' (perhaps a maverick) line of inquiry. In contrast to this generalized neglect, however – and infusing the writings comprising this volume – is a will to explore the relations between urban and cinematic space; to explore the screenscapes of the 'cinematic city'. Such an exploration is far from tangential to either film or the city, since the kind of issues raised in the chapters below are increasingly seen as vital to a proper under-standing of both. Indeed, those film and urban theorists who continue to insist on the separation of these themes might be accused of failing to appreciate vital dimensions of their respective fields – for the city has undeniably been shaped by the cinematic form, just as cinema owes much of its nature to the historical development of the city. Accordingly, in this introduction I seek to make some prefatory remarks on what Colin McArthur tellingly describes in this volume as the *elusive* cinematic city.

It is, perhaps, the writing of Walter Benjamin – whose recognition of the affinity between the city and the cinema has never been fully appreciated (cf. Natter, 1994) – to which Baudrillard's (1988) remarks on the cinematic city hark back. Influenced by ideas associated with the Viennese school of art history (see Levin, 1988) – ideas formulated, as Lant (1995) notes, contemporaneously with the development of the cinema – Benjamin (1969) expanded upon and radically revised the notion that cultural transformations register shifts in sense perception, shifts which both issue from and continue to shape fundamental historical change: 'The film corre-sponds to profound changes in the perceptive apparatus – changes that are experienced on an individual scale by the man in the street in big-city traffic' wrote Benjamin (1969: 250). The new *technik* of cinema – the German term evoking 'technology' as well as 'technique' – registered such a shift in the field of human perception, one that worked to sensitize people to aspects of the world that had previously gone unnoticed; that were not previously recognized or even recogniz-able *as* reality (Hansen, 1987). As Shaviro (1993: 41) puts it: 'Cinema is at once a form of perception and a material perceived, a new way of encountering reality and a part of reality thereby perceived for the first time.' Hence cinema's poten-tial for achieving startling new effects in its modifications of scale and proximity (for instance in the gigantism of the face in close-up); of speed and slowness (mani-fest in both the very form of film's moving pictures, described by Virilio (1988: 32) as the 'speed effect of light,' and in the process of editing and reorganizing space–time); its initial liberation of sound, of smell, and of colour from vision; and so on. 'All this is summed up in Benjamin's notion of the camera as the machine opening up the optical unconscious,' suggests Taussig (1993: 24).

It is this kind of recognition of cinema's complex relation with the real that is reiterated in Baudrillard's insistence that the cinema is an exceptional version of itself, always already leaking out of its specific context to become dispersed across a multitude of sites, most notably in the city. For 'It is there that the cinema does not assume an exceptional form, but simply invests the streets and the entire town with a mythical atmosphere' (Baudrillard, 1988: 56). If many of us have, therefore, experienced that sudden, strange feeling whilst walking in the city that we are walking through the set of a film, this is undeniably a part of the cinema. Cinema can no longer be restricted to the screen upon which films are projected, or to the darkened interior of the movie theatre where we become, directly, the spectators of film. (Though, as Schivlesbuch [1988: 221] notes, 'The power of artificial light to create its own reality only reveals itself in darkness.') And so, when David Harvey (1989: 308) writes that film is, 'in the final analysis, a spectacle projected within an enclosed space on a depthless screen,' such a foreboding 'final analysis' effects a closure that is absent from the cinema itself, which necessarily possesses the potential to leak out, continuously, all over the city; and vice versa (a theme reflexively screened in Woody Allen's *The Purple Rose of Cairo*).

(POST)MODERNITY AND THE CITY

The urbanization that accompanied the expansion of industrial capitalism was both a direct manifestation of, and itself served to shape, the historical transition towards a specifically *modern* mode of social living. Whilst both documenting and providing commentary on these developments, cultural forms such as the cinema and its various precursors were themselves implicated in such changes. Thus, the spectacle of the cinema both drew upon and contributed to the increased pace of modern city life, whilst also helping to normalize and cathect the frantic, disadjusted rhythms of the city (Crary, 1990); reflected and helped to mould the novel forms of social relations that developed in the crowded yet anonymous city streets; and both documented and helped to transform the social and physical space that the modern city represented. In a recent discussion of the social spaces of modernity, Zygmunt Bauman (1993) has sought to map the discontinuous, disjunctured spaces that were definitive of 'urbanism as a way of life' (to borrow Wirth's [1948] felicitous phrase). A concern with the social *spaces* of modernity is far from arbitrary. Modernity as a whole concerned itself with 'the harsh law of spacing' (to recontextualize a phrase from Derrida, 1976: 200) – with constructing 'intellectually, by acquisition and distribution of knowledge,' a clearly ordered, bounded and mappable 'cognitive space' (Bauman, 1993: 146). Modernity was thus marked by a powerful obsession to impose a thoroughly rationalized order on to the world, an order that would efface all traces of the ambivalence that characterized earlier modes of life. Paradoxically, however, modernity's own ordering zeal itself gave rise to the welling up, predominantly in the city, of a new embodiment of such ambivalence, captured in the fleeting figure of the *stranger*. The stranger was, in effect, the personification of all that modernity's efforts at cognitive spacing sought in vain to annihilate, and merely succeeded in displacing.

Whereas the social and physical spaces of pre-modern society formed an intimately related, lived totality, modernity brought about their colonization by a thoroughly *abstract* space, which ensured their fragmentation and disjuncture (Lefebvre, 1991). A world that was once perceived 'as a living whole', so to speak, could no longer be experienced as coherent and complete. The hallmark of the stranger, for instance, was that he or she was immediately proximate in *physical* space yet distant in *social* space. That the social and the physical worlds were no more conterminous gave rise to a new kind of *virtual* or *spectral* presence – a flickering ontology or *hauntology* (Derrida, 1994) – characteristic of the stranger. The ambivalence of the stranger thus represented the ambivalence of the modern world. Time and space were no longer stable, solid and foundational. Hence, the experience of modernity equated, in the words of Baudelaire (cited in Harvey, 1989: 10), to 'the transitory, the fleeting and the contingent'. Such was the effect of the disjuncture of both time and space that modernity inaugurated. Moreover, it was in the modern city that this sense of the ephemeral and the fragmentary made its flickering presence most intensely felt. The modern city was, concomitantly, the world as experienced by the stranger, and the experience of a world populated by strangers – a world in which a universal strangehood was coming to predominate. It was within such a world that the virtual presence of the cinema was to find its place (Friedberg, 1993; Bruno, 1993).

If, however, an increasingly institutionalized and bureaucratized modernity found itself having almost continuously to redouble its efforts in its attempt to impose a rigidly codified social order – such that it might eradicate the ambivalence continuing to 'pollute' urban society – this was a profoundly contradictory aim. As Simmel (1950, 1978) demonstrated definitively, the conditions of social life that modernity sought to impose were fundamentally dependent on precisely the kind of relation embodied in the stranger. The paradigmatic example is that of the modern monetary system: 'The desirable party for financial transactions – in which, as it has been said quite correctly, business is business – is the person completely indifferent to us, engaged neither for nor against us,' wrote Simmel (1978: 227). And so, whilst from the one vantage point the stranger seemed nothing but a potential threat, from another – ultimately inseparable from the first – the stranger is the very agency necessary for the institutional structures of modernity to function. In the face of such an aporia, other modes of coping with the modern city inevitably found the space for their development. One of the most significant finds its prototype in the figure of the *flâneur*, the celebrated nineteenth-century 'city stroller' (Benjamin, 1973a; Tester, 1994), who captures the 'convergence of new urban space, technologies and symbolic functions of images and products' (Crary, 1990: 20).

The *flâneur*'s existence was built upon the sustained disavowal of the cognitive ordering of space, in favour of a self-defined and self-centred *aesthetic* spacing.

The contours of aesthetic space were traced out by way of a ludic peregrination, etched in a dimension transversal to those efforts to impose an orderly cognitive space (cf. Deleuze and Guattari, 1988; Sennett, 1993). Hence, the *flâneur* fully embraced the uneasy, fleeting lifeworld of the modern city, enthralled by the pleasures and potentialites of a world removed from the presence, stricture and restraint of tradition, but also from the functional efficacy of modern rationality. However, as Wilson (1992) argues – *contra* Wolff (1985) – the predominantly male gendering of the nineteenth-century practice of *flânerie* signalled something of a crisis of masculinity brought about by modernity, not a simple continuation or reaffirmation of the existing phallocentric order. For the *flâneur*, the (sexually charged) pleasures of the city stemmed from an *aesthetic* proximity to others that was wholly detached from any *social* proximity (and hence from any responsibility or consequence) – summarized in Buck-Morss' (1989: 346) pithy description of the *flâneur*'s guiding principle as 'look, but don't touch'. Or, as Bauman (1993: 172) puts it, 'the city stroller can go on drawing the strangers around him into his private theatre without fear that those drawn inside will claim the rights of . . . insiders.' However, the *flâneur* arguably forged his stoical existence in the face of an overbearing perception of the inevitability of defeat (Wilson, 1992). In so far as the turbulent space of the modern city was experienced as labyrinthine and disorientating, the *flâneur*'s existence was marked by melancholic nostalgia for a lost (or impossible) world, and by a sense of impotence at the interminable deferral of any sense of arrival at a final destination.

For Buck-Morss (1989: 346), though, the *flâneur* is 'more visible in his afterlife than in his flourishing,' his living on by other means ensuring that a myriad of subsequent cultural forms 'no matter how new they appear, continue to bear his traces, as ur-form'. The aestheticized existence of the *flâneur* was, therefore, not a short-lived, temporary mode of coping with the virtuality and the mobility of the modern city. Rather, the extinction of the *flâneur* as such was only brought about by the fragmentation of the practice of *flânerie*. As Friedberg (1993: 3) suggests, for instance, '*flânerie* can be historically situated as an urban phenomenon linked to, in gradual but direct ways, the new aesthetic reception found in "moviegoing".' In particular, the practice of *flânerie* and the apparatus of the cinema both changed the social meaning of *presence*, and did so in much the same way; both effectively embraced the virtual. Similarly, the moving pictures of the cinema – and, indeed, cinema's introduction of the moving camera – shifted the nature of *mobility* itself from the sphere of the actual to that of the virtual; from movement (which carries with it a sense of direction) to the pure circulation of speed (see Kern, 1983; Schivlesbuch, 1986; Thrift, 1994). Virilio (1989: 110), for instance, refers to the introduction of the 'mobile camera' in the 1895 Lumière brothers' film *L'entrée d'un train en gare de La Ciotat* as the invention of 'the first static vehicle'.

If such modern transformations were in large measure modernity's *unintended* consequences – surprising new forms of ambivalence, far removed from modernity's own inaugurating principles – then there is more than a little weight in Friedberg's

(1993) suggestion that *flânerie* directly but gradually tore open the very fabric of modernity. European modernity initially saw and represented itself as the ultimate break from the dead weight of past tradition, custom and superstition. The modern belief in the ordering powers of Reason turned its back on customary ways of proceeding, and in so doing accomplished its release from both the authority of the Church and the legitimation of God (Chadwick, 1975). Yet, whilst the meta-narratives of modernity portrayed the progress to be gained from the institution of Reason as nothing less than the 'grand march of History', faith in History's utopian telos was to be subject to repeated questioning, despite repeated attempts to force its realization in the city – evident, for instance, in Patrick Geddes' notion of the 'eutopia' and in Le Corbusier's conception of the urban environment as comprising 'machines for living' (Fishman, 1982). The dystopian *alter ego* of moder-nity's telos was a powerful, recurrent nightmare, one that was arguably framed more forcefully in filmic representations of the city, from *Metropolis* to *Blade Runner*, than in any other form. Yet, whilst both utopian and dystopian futures were directly the product of modernity's faith in Reason and progress, it was ulti-mately the kind of transversal movement associated with both *flânerie* and the cinematic form that paved the way towards a postmodern condition – and towards the notion of heterotopia (Foucault, 1986; Genocchio, 1995; Soja, 1989).

For Lyotard (1984: xxiv), we must grasp the '*postmodern* as incredulity towards metanarratives'. The idea of History – and with it the idea of the end of History (whether utopian or dystopian; idealist or materialist) – is no longer a viable, or even an interesting, belief. We have already passed *beyond* the end – 'This is our destiny: the end of the end. We are in a transfinite universe,' writes Baudrillard (1990: 70). Utopian and dystopian futures may still preoccupy such cultural forms as film but nobody quite believes in their *reality* any more; yet they still perform an alibi-serving function or provide a deterrence effect (Baudrillard, 1994). In so far as this situation amounts to 'a crisis of historical ideals facing up to the impossibility of their real-ization' (Baudrillard, 1988: 77), it is a specifically European crisis, stemming from the extent to which European modernity is tied to its own particular conception of History. For Baudrillard (*ibid*.: 76), however, 'America is the original version of modernity. We [Europeans] are the dubbed or subtitled version.' America asserted its utopianism from the start, thus forcing a conviction in its reality – and, as Baudrillard (*ibid*: 77) adds: 'We should also not forget the fantasy consecration of this process by the cinema.' Accordingly, America's crisis is one of 'an achieved utopia, faced with the problem of its duration and permanence' (*ibid*.).

If, unlike its European counterpart, America's modernity was not based on a principle of historical accumulation that would one day yield its full and final fruits, European modernity's centralized apparatuses of power have arguably been subject to a radically decentralized and individualized postmodern transformation, reminiscent of America's own 'achieved utopia'. This transformation occurred predominantly through the *systemic* appropriation of the *flâneur*'s originally *anti-systemic* existence. The prototypical instance of such a transformation took place within the sphere of consumption. As Bauman (1993: 173) puts it: from the vantage point of capital, 'The right to look gratuitously was to be the *flâneur's*, tomorrow's *customer's*, reward.' The importance, in this process, of the specifically *visual* pleasures afforded to the would-be customer/consumer is, of course, highly significant, given that cinema and earlier technologies of visual pleasure in part 'codified and normalized the observer within rigidly defined systems of visual consumption' (Crary, 1990: 18), thus not only forming a facet of but also reflecting the continued forging of vision as 'the master sense of the modern era' (Jay, 1988: 3; see also Crary, 1988; Friedberg, 1993; on the gendering of scopophilia, see Mulvey, 1975; Penley, 1989; Rodowick, 1991). Through this transformation of modernity, the lifestyles of the inhabitants of the postmodern city are increasingly geared to the pleasures of sensation without consequence – pleasures born of modern metropolitan life and, with it, the cinema (Bruno, 1993; Friedberg, 1993). However, this is to beg the question of the precise form and the potentialities of cinema, an issue that has long been the topic of theoretical investigation.

FILM, THEORY, SPACE

Over the course of its history, film theory has demonstrated a sustained commitment to understanding the specificities of the cinema (see Andrew, 1976; Easthope, 1993; Lapsley and Westlake, 1988). Today by far the dominant paradigm for the study of film covers a theoretical terrain triangulated by semiotics, psychoanalysis and historical materialism. However, amidst fears that film theory has stultified into a new orthodoxy, a number of commentators have begun to reassess the nature of the cinematic experience. Thus, for Shaviro (1993: 36), the fascination exerted by film cannot be understood from within the dominant paradigm, which fails to grasp that cinema's *(re)production* of visual and sonoral sensation 'cannot be equated with or reduced to their *representation*'. At issue here is the tension between *representation*, which always already accords to the logic and order of the Same, and questions of *repetition*, which open on to difference and Otherness (Deleuze, 1986, 1989). Rather than according to a logic of representation, cinema works alongside the body (Shaviro, 1993), and alongside the city.

Earlier debate over the nature of the cinema, between realists such as Bazin (1967, 1971) – who celebrated cinema's power of realism over and above other art forms – and formalists, such as Arnheim (1958), Balázs (1952), and Eisenstein (1963) – who perceived cinema's vitality to lie in its ability to transcend the real – was in effect resolved and surpassed by the semiotic conception of film (see Metz, 1974a, 1974b, 1982; Wollen, 1972), upon which the current edifice of film theory stands. This development in film theory occurred as a result of the semiotic approach providing the insight that *both* sides of the formalist–realist debate rested on the misplaced assumption that cinema primarily bore an indexical relation to reality. However, just as this paradigm shift involved a recognition that a questionable common assumption was at work on both sides of the debate, for Shaviro (1993), a similar structure characterizes contemporary debate over film:

Film theorists usually posit a radical opposition between the direct presentation of sounds and images through recording and the manipulation of them through editing. This is as much the case for realists such as Bazin and contemporary phenomenological theorists such as Dudley Andrew as it is for formalists from Eisenstein to Balázs on to semiotic and psychoanalytic critics writing in the wake of Metz ... We can assert, against Bazin, that a style based on long takes and depth of field is no less constructed than one based on montage; but we can also argue, against formalists and semioticians, that this constructedness in no way compromises the perceptual intensity/ immediacy of images and movements in time and space.

(Shaviro, 1993: 35–6)

It is this sensorial immediacy, this *haptical* quality of the cinema – which detaches it from the order of the re-presentation of an original – that led Deleuze (1986, 1989) to adopt Peirce's (1991) rather than Saussure's (1974) semiotics in his cinematic studies (Ropars-Wuilleumier, 1994). And it is precisely this quality that is highlighted in Benjamin's (1969) remarks on the cinema.

However, Stephen Heath's (1981) seminal account of the 'narrative space' of the cinema – one of the remarkably few essays to engage explicitly with cinematic space – lies squarely within the semiotic–psychoanalytic tradition of film theory. Heath explicates the role that narrative has come to assume in assuring the cinema spectator a coherent positionality (see Easthope, this volume, for a lucid and succinct account of Heath's argument). For Heath, it is the case that film, whilst deploying practices of perspectival representation definitive of classical or Quattrocento space, continuously threatens to disrupt the fixity and coherence of perspectival representation, inasmuch as the cinema consists of *moving* pictures. And narrative is, on Heath's account, the dominant cinematic practice that defends against or 'contains' this disruptive mobility – thereby ensuring that the (Lacanian) subject's dissimulation of its own lack of coherence remains intact. However, whilst critiques of the 'formalism' exemplified by work such as Heath's are relatively well known – in particular its alleged 'bracketing of history' (see Jameson, 1981, 1991) – Heath's account can also be interrogated in respect of the conception of the experience undergone by the cinemagoer by considering the question of the *haptical* nature of cinematic space.

As a number of writers on early cinema have suggested (Gunning, 1981, 1983; Elsaesser with Barker, 1981; Hansen, 1991), it was only after an initial period of experimentation, technological innovation and change that narrative cinema became probably *the* dominant form. And, as Lant (1995: 47) puts it, 'the less strictly spatially coded arrangements of early cinema ... [suggest] that a wider range of spectatorial responses was possible and indeed existed' in cinema's early years. Given that the brand of film theory drawn upon and developed in Heath's (1981) account of narrative space explicitly conceives of cinema in terms of its visuality rather than any hapticality, to what extent might film theory be complicit with this development in cinema itself? And to what extent does it adequately grasp the actuality – or, rather, the virtuality – of the cinematic experience?

For Benjamin, cinema represented a form that operated in a manner 'primarily tactile, being based on changes of place and focus which periodically assail the spectator' (Benjamin, 1969: 238). It is here that Benjamin's relation to such theories as those of Adolf Hildebrand (1978 [orig. 1893]) and, especially, Aloïs Riegl (1993 [orig. 1893]; 1985 [orig. 1901]) – theories which amounted to comparative attempts to conceptualize the distinctiveness of the cultures of different epochs and societies – becomes evident. Such works provided a conceptual apparatus for considering art that entailed a non-perspectival space, with Ancient Egyptian art being the key example. Wilhelm Worringer, for instance – himself greatly influenced by Riegl – saw 'the problem of the history of the genesis of space' as 'a truly modern one' (1928 [orig. 1927]: 73), and held that Egyptian art was particularly instructive inasmuch as it represented 'the pronounced counterpart of our view of space' (ibid.: 82). Egyptian art, which many contemporary commentators – though Benjamin was not amongst them – saw as curiously resonant with cinema (Lant, 1992), required the engagement of the subject in close proximity – Nahsicht, or near-sight, in Hildebrand's terms – where the eye engages in a haptical rather than simply optical mode of perception. In Deleuze's (1981: 99) words, such a situation is where 'sight discovers in itself a function of touching that belongs to it and to it alone and which is independent of its optical function' (cited in Polan, 1994: 252). It is in accordance with this situation that cinematic space cannot be simply equated with a perspectival representation of (another) space, its dynamism contained by its narrative form (thereby offering the spectator a seemingly coherent position of phallocentric visual mastery).

The hapticality of the cinema pertains to the extent to which film is, unlike painting when it employs perspectival representation, founded in mechanical (re)production – that is, to the extent that it is simulacral rather than representational. Benjamin's remarks on the difference between theatre and film bring to centre stage the question of the relation between (re)production, representation and the cinemagoer. 'The stage actor identifies himself with the character of his role. The film actor very often is denied this opportunity,' writes Benjamin (1969: 230), underscoring that 'For . . . film, what matters is that the actor represents himself to the public before the camera' (ibid.: 229). And, given that the cinemagoer occupies the vantage point of the static vehicle par excellence, the movie camera itself, then rather than a representation of space as such, film (re)produces a virtual space, marked by a 'proximity without presence' (Fleisch, 1987: 44). Such a space offers a completely different mode of engagement than that theorized by Heath (1981). 'What happens', asks Blanchot (1981: 75), 'when what you see, even though from a distance, seems to touch you with a grasping contact, when the matter of seeing is a sort of touch, when seeing is a contact at a distance?' The answer, of profound relevance for the cinemagoer, is the experience of a certain captivation and fascination. The haptical space of the cinema potentially serves to transform the division between the Eye (= I) of the Rational Subject and its (assimilable and rationalizable) object – the very division definitive of the frame of reference through which modernity has imagined itself: 'The split, which had been the possibility of seeing, solidifies, right inside the gaze, into impossibility. In this way, in the very thing that makes it possible, the gaze finds the power that neutralizes it' (Deleuze, 1986: 75). In achieving such a re-framing of the world, the camera's penetration of reality entails a transformation in the perception of the cinemagoer, and does so in a manner consonant with the experiences offered by the flickering, virtual presence of the city (Friedberg, 1993).

THE CINEMATIC CITY: TAKE 2

The aim of this volume is to provide for the *exploration* of the cinematic city without imposing, a priori, any single interpretation. The many differences between these prefatory remarks on the city and the cinema, and the varied approaches taken up in subsequent chapters, will, I hope, serve to open up a space for the continued theorization of the cinematic city. None the less, the organization imposed here does follows a particular rationale. Taken together, this introduction and the chapter by Colin McArthur immediately following it should provide sufficient orientation for the reader to encounter the many facets of the cinematic city discussed in the remainder of the book. In Chapter 1, McArthur's endeavour to track the elusive cinematic city highlights, particularly, the structural opposition between city and country, an opposition that has been mobilized in an immense variety of ways over the course of film history. Thus, ranging from early to postmodern cinema, across European, American and Third World cinema, and through genres as diverse as the musical and the gangster film, McArthur provides a panoramic journey through both the mainstreets and byways of the cinematic city. This journey provides for a detailed discussion of what will become a recurrent theme in subsequent chapters: the figuration of the city in cinema as either utopian or dystopian, and its relation to the extra-cinematic city.

In Chapters 2–8 the cinematic city is considered through the analysis of a range of filmic texts, genres and contexts, from which a variety of theoretical considerations emerge. In Chapter 2, Giuliana Bruno's 'City Views' provides a detailed and illuminating study of the early Neapolitan film industry and the 'voyage of film images' established between Italian and US cities. This essay complements her 1993 book-length study of the city films of Elvira Notari, *Streetwalking on a Ruined Map*. There, Bruno provides a 'microhistorical study' of a forgotten world of film, which she describes as a kind of confrontation 'with a ruined and fragmentary map', which folds out into a full-scale consideration of the contours of modernity. In the essay presented here, Bruno continues this project of cultural mapping, pursuing the diasporic urban history of both films and peoples in a discussion that is historically grounded, yet encompasses the many and complex dimensions of modernity as a whole.

Whilst Bruno's essay emphasizes the transience, movement and flows characteristic of modernity's conjunction with the city and the cinema – echoing Baudelaire's description of modernity as 'the transient, the fleeting, the contingent' – in Chapter

3, the analysis by John Gold and Stephen Ward of British documentary film brings out the other half of modernity; 'the eternal and the immutable' (Baudelaire, cited in Harvey, 1989: 10). Or, more precisely, their discussion of those documentary films dealing with the planning of the future city reveals modernity's attempt to *realize* the 'eternal' and 'immutable'. Elsewhere, for instance, John Gold (1985) has noted the role of the Italian Futurist movement in presaging the main dimensions of virtually all subsequent modernist visions of the future city. In the chapter included here, Gold and Ward provide a detailed account of a specific historical attempt to *realize* utopian modern urban visions in British town planning, showing the way in which documentary film was mobilized in an effort to popularize, universalize and above all legitimate such visions (see also Gold and Ward, 1994).

If Gold and Ward's chapter analyses the sustained efforts to project – both cinematically and in the built form of cities – a utopian urban future, in 'Something More than Night: tales of the *noir* city' (Chapter 4), Frank Krutnik demonstrates the way in which one particular film genre has engaged with the shadowy, dystopian world of the city (see also Krutnik, 1991). In this chapter, Krutnik provides an exemplary discussion of the sense of place conjured by film, detailing how *film noir*'s engagement with the city produces a profound sense of dis-location, a sense that is also present as a narrative device in such films as Frank Capra's *It's a Wonderful Life* (1946). As Krutnik demonstrates, the darkness, crime, threat and desire to which the protagonists of the *noir* city are subject trace out a particularly powerful mode of imagining the city – one that recalls Wilson's (1992) discussion of the longing and impotence first experienced by the nineteenth-century *flâneur*. Hence, whereas the juxtaposition of the chapters by Bruno and by Gold and Ward brings out the contradictory dimensions of modernity as a whole, the juxtaposition of the chapters by Gold and Ward and by Krutnik highlights a fundamental contradiction immanent to the modern city itself. Taken together, these latter two chapters condense and crystallize the all-or-nothing modernist image of the city as *either* utopia *or* dystopia.

Whilst *film noir* represents a relatively familiar genre, in Chapter 5 Will Straw considers a far less familiar period of film history, which might be described as one of *film noir*'s mutant off-shoots. Identifying a cycle of 'urban confidential' and 'exposé' films – which almost wallow in the new depths of degradation and criminality that are paranoiacally assumed to be rife in the city – Straw offers a fascinating insight into the sensibilities of American popular culture of the 1950s. His discussion of the 'lurid city' traces not only its portrayal in film but also its discursive construction across a variety of other social and cultural forms, from pulp journalism to governmental inquiries. If Straw's chapter might be regarded as conveying something of a crisis point in relation to the trajectory of modernity – figured as a kind of turning in on itself – then Antony Easthope's focus on some of the 'Cinécities in the Sixties' (Chapter 6) signals the untimely arrival of the postmodern. Beginning with a recapitulation of the utopian and dystopian representations of the modern city explored in earlier chapters, and a discussion of Foucault's conceptualization of the heterotopia, Easthope provides an engaging look at the

questions of modernity and postmodernity raised by visions of the city in, particularly, the films of Godard and Antonioni. And, in so far as the order of the essays
presented thus far has followed a loosely historical sequencing, Easthope's chapter
offers an explicit engagement with the question of the modern and the postmodern
that is potentially disruptive of such a history – an issue that is returned to in the
next two chapters of the book.

The spectre of the postmodern raised in Easthope's chapter is immediately
broached again in the reconsideration by Marcus Doel and David Clarke, in
Chapter 7, of *Blade Runner* – a film that has already been marked out, by an
extensive body of literature, as an exemplary 'postmodern' film offering the ultimate
dystopian urban vision. Doel and Clarke, however, question such a reading of the
film which, they demonstrate, amounts to a thoroughly modern one. Focusing on
the film's deployment of the Eye as a central motif, they argue against viewing
cinema as a mirror. Instead, they seek to uncover its contradictory screening of a
modernist impulse towards order and the heterology of 'symbolic exchange' that
continues to haunt this ordering impulse, particularly through the reversibility of
life in death. This interrogation of the postmodern continues in Chapter 8, where
Elisabeth Mahoney raises some serious questions regarding the *politics* of postmodernism. Through readings of *Night on Earth*, *Falling Down*, and *Just Another
Girl on the I.R.T.*, Mahoney engages with both recent feminist theory and critical
human geography in order to demonstrate the evident refusal of 'modern', political concerns – particularly those based around race, gender and identity – simply
to disappear in supposedly 'postmodern' films, times and spaces. Whilst drawing
on different theoretical perspectives, therefore, the chapters by Easthope, Doel and
Clarke, and Mahoney each serve to raise a number of important questions regarding
the postmodern.

The chapters discussed thus far generate a variety of theoretical insights primarily
from their engagement with particular films, genres or contexts. The final three
chapters of the book each seek to make some broader observations on the nature
of the cinematic city. In Chapter 9, Rob Lapsley provides a detailed theoretical
reflection on the role that the city has played in film. As if to challenge those who
would wish to banish the semiotic–psychoanalytic conception of film theory, Lapsley
deploys a virtuosic grasp of Lacanian theory in order to provide an insightful,
sensitive and yet uncompromising discussion – which demonstrates that such a
desire might not be quite so straightforward. If, as noted above, a number of
writers have marshalled Benjamin to urge a film–theoretical re-take, Lapsley's
demonstration of the continuing significance of Lacanian theoretical formulations
will prompt a reassessment of many recent negative pronouncements on the state
of health of the semiotic–psychoanalytic film paradigm. Nevertheless, in Chapter
10, James Hay provides an equally incisive and insistent theoretical critique of the
issues that film theory of this kind has allegedly overlooked.

Hay's concerns are pointed up, initially, by looking at the question of film history,
which he links both to celebrations of cinema's centennial anniversary in 1995
and to academic film theory's historicism. The latter he attempts to remedy by a

reconsideration of the relations between cinema and space. In so doing, he shares something of Jameson's (1992) concerns, though Hay avoids Jameson's argument as scrupulously as he distances himself from the difficulties he identifies in Heath's (1981) conception of 'narrative space' and Soja's (1989) *Postmodern Geographies*. Drawing initially on the case of Italian film between the two World Wars and its relation to the emergence of a national popular culture (cf. Hay, 1987), and finally broadening out his discussion to consider the present day Hollywood system and the media economies of other US cities, Hay develops a conception of the cinematic that is sensitized to cinema's broader relation to the city, and beyond. The chapters by Lapsley and Hay together, therefore, offer an exceptionally clear crystallization of some of the most important, contested dimensions of current film theory. Of course, film is just one cultural form, which reveals its particular affinity with the city in a number of fundamental ways. As Silverman (1988) and Chon (1994) have argued in different ways, however, cinema possesses significant *aural* as well as visual effectivities. In the final chapter of the book (Chapter 11), therefore, Iain Chambers takes film as a point of departure from which to frame some broader considerations of city culture, music and memory. His insistent reference to music permits him to defer the question of the cinematic city *per se*, and hence to move towards the composition of other, different questions of urban culture, space and place. It is, therefore, by way of an opening out from the cinematic city and on to a far broader urban cultural terrain that this book finds a conclusion – but not a closure.

If the conception of the cinematic city is capable of generating such a range of complex and valuable insights into both cities and film, this demonstrates the pertinence of these two terms to any consideration of (post)modernity. Film theory has frequently been regarded as being in advance of most other accounts of aesthetic texts. Considerations of space, spatiality and geography have more recently come to the fore in critical attempts to grasp the changing shape of the social world. The benefits of bringing together these modes of understanding are in many ways taken as axiomatic, though certainly not as unproblematic, in the selection of essays gathered together here. The practice of such a task remains, however, a continuing challenge.

ACKNOWLEDGEMENTS

Thanks to Marcus Doel, Martin Purvis and Paul Waley for their detailed comments on this introduction. It remains to acknowledge the debts I have incurred in bringing this work as a whole to completion: thanks, first and foremost, to all the contributors to this volume for all their efforts; to Janice Campbell, former Director of the Leeds International Film Festival, whose enthusiasm and support permitted the 1993 *Screenscapes* conference to take place, out of which the idea for this volume initially sprang; to Mike Bradford and Antony Easthope for editorial suggestions; to the British Film Institute for help with providing the majority of the illustrations;

and, last but not least, to Tristan Palmer, Matthew Smith and Katherine Hodkinson at Routledge for their help and guidance in bringing this volume to fruition. While every effort has been made to trace copyright holders and obtain permission, this has not been possible in all cases. Any omissions brought to our attention will be remedied at the earliest opportunity. In keeping with my editorial responsibilities I should add, however, that I alone accept responsibility for any shortcomings of the volume as a whole.

REFERENCES

Aitken, S. C. and Zonn, L. E. (eds) (1994) *Place, Power, Situation, and Spectacle: A Geography of Film*, Lanham, Md, Rowman and Littlefield.

Andrew, D. (1976) *The Major Film Theories: An Introduction*, Oxford, Oxford University Press.

Arnheim, R. (1958) *Film as Art*, London, Faber.

Arscott, C. and Pollock, G. (1988) 'The partial view: the visual representation of the early nineteenth-century industrial city', in J. Wolff and J. Seed (eds) *The Culture of Capital: Art, Power and the Nineteenth-century Middle Class*, Manchester, Manchester University Press, 227–9.

Balázs, B. (1952) *Theory of the Film: Character and Growth of a New Art*, London, Dobson.

Batchen, G. (1991) 'Desiring production itself: notes on the invention of photography', in R. Diprose and R. Ferrell (eds) *Cartographies*, Sydney, Allen and Unwin, 13–26.

Baudrillard, J. (1988) *America*, London, Verso.

Baudrillard, J. (1990) *Fatal Strategies*, London, Pluto Press.

Baudrillard, J. (1994) *Simulacra and Simulation*, Ann Arbor, University of Michigan Press.

Bauman, Z. (1993) *Postmodern Ethics*, Oxford, Blackwell.

Bauman, Z. (1995) *Life in Fragments: Essays in Postmodern Morality*, Oxford, Blackwell.

Bazin, A. (1967) *What is Cinema? Volume 1*, Berkeley, University of California Press.

Bazin, A. (1971) *What is Cinema? Volume 2*, Berkeley, University of California Press.

Benjamin, W. (1969) 'The work of art in the age of mechanical reproduction', in *Illuminations*, New York, Schocken, 217–51.

Benjamin, W. (1973) *Charles Baudelaire: A Lyric Poet in the Era of High Capitalism*, London, New Left Books.

Blanchot, M. (1981) *The Gaze of Orpheus*, New York, Station Hill.

Bruno, G. (1993) *Streetwalking on a Ruined Map: Cultural Theory and the City Films of Elvira Notari*, Princeton, NJ, Princeton University Press.

Buck-Morss, S. (1989) *The Dialectics of Seeing: Walter Benjamin and the Arcades Project*, Cambridge, Mass., MIT Press.

Chadwick, O. (1975) *The Secularization of the European Mind in the Nineteenth Century*, Cambridge, Cambridge University Press.

Chon, M. (1994) *Audio-Vision: Sound on Screen*, New York, Columbia University Press.

Clark, T. J. (1984) *The Painting of Modern Life: Paris in the Art of Manet and His Followers*, London, Thames and Hudson.

Clarke, D. B. (1994) 'Commentary: film/space', *Environment and Planning A* 26, 1821–23.

Crary, J. (1988) 'Modernizing vision' in H. Foster (ed.) *Vision and Visuality*, Seattle, Wash., Bay Press, 29–49.

Crary, J. (1990) *Techniques of the Observer: On Vision and Modernity in the Nineteenth Century*, Cambridge, Mass., MIT Press.

Daniels, S. (1986) 'The implications of industry: Turner and Leeds', *Turner Studies* 6, 100–17.

Deleuze, G. (1981) *Logique de la Sensation*, Paris, Editions de la différence.

Deleuze, G. (1986) *Cinema 1: The Movement-Image*, London, Athlone.

Deleuze, G. (1989) *Cinema 2: The Time-Image*, London, Athlone.

Deleuze, G. and Guattari, F. (1988) *A Thousand Plateaus: Capitalism and Schizophrenia*, London, Athlone.

Derrida, J. (1976) *Of Grammatology*, Baltimore, Md, Johns Hopkins University Press.

Derrida, J. (1994) *Spectres of Marx: The Work of Mourning, the State of the Debt, and the New International*, London, Routledge.

Easthope, A. (ed.) (1993) *Contemporary Film Theory*, London, Longman.

Eisenstein, S. (1963) *Film Form: Essays in Film Theory*, London, Dobson.

Elsaesser, T. with Barker, A. (eds) (1981) *Early Cinema: Space, Frame, Narrative*, London, British Film Institute.

Fishman, R. (1982) *Urban Utopias in the Twentieth Century*, Cambridge, Mass., Harvard University Press.

Fleisch, W. (1987) 'Proximity and power: Shakespearean and cinematic space', *Theatre Journal* 4, 277–93.

Foucault, M. (1986) 'Of Other Spaces', *Diacritics* 16(1), 22–7.

Friedberg, A. (1993) *Window Shopping: Cinema and the Postmodern*, Berkeley, University of California Press.

Genocchio, B. (1995) 'Discourse, discontinuity, difference: the question of "other" spaces', in K. Gibson and S. Watson (eds) *Postmodern Cities and Spaces*, Oxford, Blackwell, 35–46.

Gold, J. R. (1984) *The City in Film*, Architecture Series 1218, Monticello, Ill., Vance Bibliographies.

Gold, J. R. (1985) 'From "Metropolis" to "The City": film visions of the future city, 1919–39', in J. Burgess and J. R. Gold (eds) *Geography, the Media and Popular Culture*, London, Croom Helm, 123–43.

Gold, J. R. and Ward, S. V. (1994) '"We're going to do it right this time": cinematic representations of urban planning and the British New Towns, 1939–1951', in S. C. Aitken and L. E. Zonn (eds) *Place, Power, Situation, and Spectacle: A Geography of Film*, Lanham, Md, Rowman and Littlefield, 229–58.

Gunning, T. (1981) 'The cinema of attractions: early film, its spectators and the avant-garde', in T. Elsaesser with A. Barker (eds) *Early Cinema: Space, Frame, Narrative*, London, British Film Institute, 56–67.

Gunning, T. (1983) 'An unseen energy swallows space: the space in early film and its relation to American Avant-Garde film', in J. Fell (ed.) *Film Before Griffith*, Berkeley, University of California Press, 356–70.

Hansen, M. (1987) 'Benjamin, cinema and experience: "the blue flower in the land of technology"', *New German Critique* 40, 179–224.

Hansen, M. (1991) *Babel and Babylon: Spectatorship in American Silent Film*, Cambridge, Mass., Harvard University Press.

Harvey, D. (1989) *The Condition of Postmodernity: An Enquiry into the Origins of Cultural Change*, Oxford, Blackwell.

Hay, J. (1987) *Popular Film Culture in Fascist Italy: The Passing of the Rex*, Bloomington, Indiana University Press.

Heath, S. (1981) 'Narrative space', in *Questions of Cinema*, London, Macmillan, 19–75.

Hildebrand, A. (1978) *The Problem of Form in Painting and Sculpture*, New York, Garland.

Jameson, F. (1981) *The Political Unconscious: Narrative as Socially Symbolic Act*, Ithaca, NY, Cornell University Press.

Jameson, F. (1991) *Signatures of the Visible*, London, Routledge.

Jameson, F. (1992) *The Geopolitical Aesthetic: Cinema and Space in the World System*, London, British Film Institute.

Jay, M. (1988) 'Scopic regimes of modernity' in H. Foster (ed.) *Vision and Visuality*, Seattle, Wash., Bay Press, 3–23.

Kern, S. (1983) *The Culture of Time and Space, 1880–1918*, Cambridge, Mass., Harvard University Press.

Krutnik, F. (1991) *In a Lonely Street: Film Noir, Genre, Masculinity*, London, Routledge.

Lant, A. (1992) 'The curse of the pharaoh, or how cinema contracted Egyptomania', *October* 59, 86–112.

Lant, A. (1995) 'Haptical cinema', *October* 74, 45–73.

Lapsley, R. and Westlake, M. (1988) *Film Theory: An Introduction*, Manchester, Manchester University Press.

Lefebvre, H. (1991) *The Production of Space*, Oxford, Blackwell.

Levin, T. Y. (1988) 'Walter Benjamin and the theory of art history', *October* 47, 77–83.

Lyotard, J.-F. (1984) *The Postmodern Condition: A Report on Knowledge*, Manchester, Manchester University Press.

Metz, C. (1974a) *Language and Cinema*, The Hague, Mouton.

Metz, C. (1974b) *Film Language: A Semiotics of Cinema*, Oxford, Oxford University Press.

Metz, C. (1982) *Psychoanalysis and Cinema: The Imaginary Signifier*, London, Macmillan.

Mulvey, L. (1975) 'Visual pleasure and narrative cinema', *Screen* 16 (3), 6–18.

Natter, W. (1994) 'The city as cinematic space: modernism and place in *Berlin, Symphony of a City*', in S. C. Aitken and L. E. Zonn (eds) *Place, Power, Situation, and Spectacle: A Geography of Film*, Lanham, Md, Rowman and Littlefield, 203–27.

Olsen, D. (1983) 'The city as a work of art', in D. Fraser and A. Sutcliffe (eds) *The Pursuit of Urban History*, London, Edward Arnold, 264–85.

Penley, C. (1989) *The Future of an Illusion: Film, Feminism and Psychoanalysis*, Minneapolis, University of Minnesota Press.

Peirce, C. S. (1991) *Peirce on Signs: Writings on Semiotics by Charles Sanders Peirce*, Chapel Hill, University of North Carolina Press.

Polan, D. (1994) 'Francis Bacon: the logic of sensation', in C. V. Boundas and D. Olkowski (eds) *Gilles Deleuze and the Theatre of Philosophy*, London, Routledge, 229–54.

Pollock, G. (1988) *Vision and Difference: Femininity, Feminism and the Histories of Art*, London, Routledge.

Riegl, A. (1985) *Problems of Style: Foundations for a History of Ornament*, Princeton, NJ, Princeton University Press.

Riegl, A. (1993) *Late Roman Art Industry*, Rome, Giorgio Bretschneider Editore.

Rodowick, D. N. (1991) *The Difficulty of Difference: Psychoanalysis, Sexual Difference and Film Theory*, London, Routledge.

Ropars-Wuilleumier, M.-C. (1994) 'The cinema, reader of Gilles Deleuze', in C. V. Boundas and D. Olkowski (eds) *Gilles Deleuze and the Theatre of Philosophy*, London, Routledge, 255–61.

Saussure, F. de (1974) *Course in General Linguistics*, London, Fontana.

Schivlesbuch, W. (1986) *The Railway Journey: The Industrialisation of Time and Space in the Nineteenth Century*, Berkeley, University of California Press.

Schivlesbuch, W. (1988) *Disenchanted Night: The Industrialisation of Light in the Nineteenth Century*, Berkeley, University of California Press.

Sennett, R. (1993) *The Conscience of the Eye: The Design and Social Life of Cities*, London, Faber and Faber.

Shapiro, T. (1984) 'The metropolis in the visual arts: Paris, Berlin, New York, 1890–1940', in A. Sutcliffe (ed.) *Metropolis: 1890–1940*, London, Mansell, 95–128.

Shaviro, S. (1993) *The Cinematic Body*, Minneapolis, University of Minnesota Press.

Silverman, K. (1988) *The Acoustic Mirror: The Female Voice in Psychoanalysis and Cinema*, Bloomington, Indiana University Press.

Simmel, G. (1950) 'The metropolis and mental life', in K. Wolff (ed.) *The Sociology of Georg Simmel*, Glencoe, Ill, Free Press, 409–24.

Simmel, G. (1978) *The Philosophy of Money*, London, Routledge.

Soja, E. (1989) *Postmodern Geographies: The Reassertion of Space in Critical Social Theory*, London, Verso.

Sutcliffe, A. (1984) 'The metropolis in the cinema', in A. Sutcliffe (ed.) *Metropolis, 1890–1940*, London, Mansell, 147–71.

Tagg, J. (1988) *The Burden of Representation: Essays on Photographies and Histories*, London, Macmillan.

Tagg, J. (1990) 'The discontinuous city: picturing and the discursive field', *Strategies* 3, 138–58.

Taussig, M. (1993) *Mimesis and Alterity: A Particular History of the Senses*, London, Routledge.

Tester, K. (ed.) (1994) *The Flâneur*, London, Routledge.

Thrift, N. (1994) 'Inhuman geographies: landscapes of speed, light and power', in P. Cloke, M. Doel, D. Matless, M. Phillips and N. Thrift *Writing the Rural: Five Cultural Geographies*, London, Paul Chapman, 191–248.

Virilio, P. (1988) *La Machine de Vision*, Paris, Galilée.

Virilio, P. (1989) 'The last vehicle', in D. Kamper and C. Wulf (eds) *Looking Back on the End of the World*, New York, Semiotext(e), 106–19.

Wilson, E. (1992) 'The invisible *flâneur*', *New Left Review* 191, 90–110.

Wirth, L. (1948) 'Urbanism as a way of life', *American Journal of Sociology* 4, 1–24.

Wolff, J. (1985) 'The invisible *flâneuse*: women and the literature of modernity', *Theory, Culture and Society* 2 (3), 37–46.

Wollen, P. (1972) *Signs and Meaning in the Cinema*, London, Secker and Warburg/BFI.

Worringer, W. (1928) *Egyptian Art*, London, Putnam.

1

CHINESE BOXES AND RUSSIAN DOLLS
tracking the elusive cinematic city
•

Colin McArthur

Alasdair Gray's novel *Lanark* contains the following much-quoted passage:

> 'Glasgow is a magnificent city,' said Thaw.' . . . Think of Florence, Paris, London, New York. Nobody visiting them for the first time is a stranger, because he's already visited them in paintings, novels, history books and films. But if a city hasn't been used by an artist not even the inhabitants live there imaginatively.'
>
> (Gray, 1981: 243)

Thaw is both right and wrong. He is right to emphasise *art* as a key domain within which sense of place is articulated, but wrong to limit this process solely to artists. Gray implicitly recognises this by including 'history books' in Thaw's list, but this is undermined once more by his insisting that the artist is the sole operational figure in the process. The key missing concept is, of course, *discourse*, a concept which, as well as including works of art, would also encompass more obviously 'factual' productions such as official reports, newspaper accounts, photographs, anecdotes, and jokes. This is not to argue that there are no pertinent differences between works of art and these other sites, simply that they might all be considered under the rubric of *discourse*. From this point of view, *pace* Thaw, Glasgow has a substantial discursive presence which has been influential in determining popular attitudes to that city and shaping subsequent narratives about it.

It is perhaps surprising that, in eighteenth century travellers' accounts, Glasgow is most often compared with Oxford for the beauty of its prospect and the excellence of its ambience. It is post-Industrial Revolution accounts of the city which begin to articulate the 'Glasgow discourse' which was to become hegemonic. Initially signalled in urban planning and public health reports in the nineteenth century, this discourse was powerfully accelerated by tabloid journalist accounts of gang warfare in inter-war Glasgow and by folkloric embellishments of these, with the result that a monstrous Ur-Narrative comes into play when anyone (not least, it should be said, Glaswegians themselves) seeks to describe or deal imaginatively with that city. In this archetypal story, Glasgow is the City of Dreadful Night with the worst slums in Europe, infested with what one English novelist has called 'a malignantly ugly people' living out lives which are nasty, brutish and short. The milieu of Glasgow is so stark, the story runs, that it breeds a particular social type, the Hard Man, whose universe is bounded by football, heavy drinking and (often sectarian) violence. This image, which beckons Circe-like to any who would speak or write of Glasgow, is about men celebrating, coming to terms with or (rarely) transcending their bleak milieu. An order of marginalisation, if not exclusion, is served on women. Mothers or lovers of the hard-eyed men who stalk this world, they are allowed no space for their own sense of what it means to be a woman in Glasgow. So all-embracing is this narrative that it can provide the underlying *langue* for the jocular act of *parole* in which a Glaswegian announces in Edinburgh (the Athens of the North) that he himself is from 'the Sparta of the North'. One wonders too how active the narrative was in producing the (perhaps apocryphal) medical statistic that the unhealthiest man in Britain is likely to be a middle-aged, bachelor barman who smokes and lives in Glasgow.

This, then, is the hegemonic Glasgow narrative but, like many hegemonies, it is fragile and contested. Other narratives about Glasgow circulate: Glasgow as the centre of 'the Red Clyde', the region where communist revolution in these islands was thought most likely; Glasgow as the warm, open-hearted city that welcomes all-comers (the implicit contrast here is with cold, thin-lipped Edinburgh); and, the most recent arrival, Glasgow as City of Culture. This latter is distinguished from the earlier narratives by the substantial presence of public relations consultants in its making, initially under the rubric 'Glasgow's Miles Better', a slogan which implicitly refers to the discourse it seeks to supercede – Glasgow as City of Dreadful Night (Spring, 1990).

What the example of Glasgow indicates is that cities (and, indeed, all urban spaces and even 'natural' landscapes) are always already social and ideological, immersed in narrative, constantly moving chess pieces in the game of defining and redefining utopias and dystopias. It should come as no surprise, therefore, that cities in discourse have no absolute and fixed meaning, only a temporary, positional one. As the opposition Edinburgh/Athens: Glasgow/Sparta shows, to cite one half of the dualism is to invoke the other, even if it is not explicitly mentioned. The high ideological valency of the discursive city and its volatility of meaning has meant that it has been diversely mobilised within the great transitions of history: the rural to the urban; the agrarian to the industrial; and – the big one – the feudal to the capitalist.[1] The lack of fixedness of meaning cannot be overstressed. Within those great historical oppositions which shaped the modern world a whole range of spaces can be identified: wildernesses; pastoral landscapes; agrarian landscapes; villages; rural towns; suburbs; inner cities; metropolises. Several diverse binary oppositions from among this list may be deployed to signify the discursive positions characteristically taken within the great historical debates. Thus, in one version, the wilderness may be set against all urban spaces as a sign of moral worth, while in another version the small town might be valorised at the expense of the inner city. In considering the question of cities in films, therefore, one has to be alert to such changing valencies and to the possibility that while, in certain representations, a structural opposition to the city may be present in the text in question, in others it may be only implicit or even wholly absent.

It has been suggested (Ford, 1994) that in American[2] silent cinema, roughly pre-1930, cities were used as random backdrops for action, the implication being that the city milieu in these films carried little or no ideological charge, so to speak, that the films were saying nothing in particular about cities. This argument is perhaps tenable if applied only to such films as Keystone cops and Harold Lloyd comedies, but alongside these were other films in which the debates around the transition to modernity, and the place of the city within these, resonated profoundly. The debate could not have been more starkly stated than in the opening of *Lights of New York* (1928). The first shot of a rural townscape is accompanied by the intertitle 'Main Street. Forty-five minutes from Broadway – but a thousand miles away'; and a later shot of the city carries the intertitle 'Broadway. Forty-five minutes from Main Street – but a *million* miles away'. This stark opposition is played out dramatically in the film. The country/city opposition, with the former as Arcadia and the latter as Sodom, furnishes the key structural opposition in *Sunrise* (1927). The plot is very basic: a woman from the city comes to the country and desta-bilises the marriage of a country couple. The master opposition of country/city is so deeply woven into the fabric of the film that a whole series of sub-oppositions can be laid out which contrast the city girl with the country girl:

city girl	*country girl*
dark, short hair	long, blonde hair
short dress	long dress
smoking	non-smoking
made-up	not made-up
unmarried	married
undomesticated	domesticated
erotic	chaste
etc.	*etc.*

So deeply grounded in (American) culture is the country/city opposition that it surfaces in the least expected places. For instance, in the film musical *The Barkeleys of Broadway* (1949), the central players, Fred Astaire and Ginger Rogers, decked out in sports clothes, disembark enthusiastically from a train for a weekend in the country. However, they have dragged along unwillingly their friend Oscar Levant – in life and art a quintessentially urban neurotic – still in his city clothes. As Oscar is frogmarched along a country lane by the others, the three of them launch into 'A Weekend in the Country', a kind of question and answer song which starkly states and restates the country/city opposition, Astaire and Rogers singing the praises of the country, Levant of the city:

F.A.	WITH GOLF AND TENNIS ROUND YOU AND NO CARES TO HOUND YOU
G.R.	WHEN MOTHER NATURE BECKONS WHO CAN DECLINE?
O.L.	TILL MOTHER NATURE VETOES THE BEES AND MOSQUITOES, MOTHER NATURE IS NO MOTHER OF MINE.
F.A./G.R.	FROM SATURDAY NIGHT TO MONDAY MORN THERE'S ALWAYS JOY AHEAD.
O.L.	FROM SATURDAY NIGHT TO MONDAY MORN I WISH THAT I WAS DEAD.
F.A./G.R.	A WEEKEND IN THE COUNTRY WILL NEVER LET YOU DOWN.
O.L.	YOU'LL PARDON MY EFFRONT'RY, BUT I'D RATHER SPEND IT IN TOWN.
F.A./G.R.	A WEEKEND IN THE COUNTRY, HEALTHY AND FULL OF SPORT,
G.R.	AND THEN THERE IS THE SALT POTATOES . . .
F.A.	WHEN YOU GET THE FRESH TOMATOES.
O.L.	I'VE GOT A LIST OF FRESH TOMATOES SUING ME NOW IN COURT.
G.R.	OH, GIVE ME THE MILK FROM THE MOO COW.
F.A.	OF CORN, RIGHT FROM THE FIELD, I'M FOND.
O.L.	IN TOWN I'D BE SPLURGIN' ON VENISON AND STURGEON BESIDE A BEAUTIFUL BLONDE.
F.A./G.R.	A WEEK CAN GET YOU SUNBURNED. VITAMIN A YOU'LL WIN.
O.L.	I'D RATHER GET HOME UNBURNED WITH MY ORIGINAL SKIN.
F.A./G.R.	A WEEKEND IN THE COUNTRY, GLORIOUS NO DOUBT.
O.L.	A WEEKEND IN THE COUNTRY? WHEN'S THE NEXT TRAIN OUT?
F.A./G.R.	A WEEKEND IN THE COUNTRY, TREES IN THE ORCHARD CALL.
O.L.	WHEN YOU'VE EXAMINED ONE TREE YOU'VE EXAMINED THEM ALL.
F.A./G.R.	A WEEKEND IN THE COUNTRY HAPPILY WE ENDORSE.
G.R.	COME GET YOUR SHARE OF NATURE'S BOUNTY.
F.A.	RIDE THE TRAIL AROUND THE COUNTY.
O.L.	I AM NO CANADIAN MOUNTIE. WHY DO I NEED A HORSE?
G.R.	HARK, HARK TO THE SONG OF THE BULLFROG.
F.A.	AT DAWN YOU RISE TO THE SONG OF THE LARK.
O.L.	WHEN ROOSTERS RIOT, I MUCH PREFER THE QUIET OF FORTY-SECOND AND FOURTH.
F.A./G.R.	GET HAPPY AND ALIVE-EE! DON'T BE A CITY POLK.
O.L.	I ONCE GOT POISON IVY.
G.R.	WILL YOU TRY FOR POISON OAK?
F.A.	A WEEKEND IN THE COUNTRY . . .
G.R.	DICKY-BIRDS OVERHEAD.
O.L.	A WEEKEND IN THE COUNTRY? I SHOULD HAVE STAYED IN BED.

In *Sunrise* the country/city opposition is played out largely in the country with one brief visit to the city. In *Mr Deeds Goes to Town* (1936), however, the same opposition is played out in the city. Longfellow Deeds (Gary Cooper), a simple, down to earth resident of a small town in rural Vermont, inherits twenty million dollars and is precipitated into the millionaire's life of the city. When he attempts to give the money away to Depression-hit American working men in the form of agricultural smallholdings, the slick city lawyers administering the inheritance attempt to have him declared insane. As Sam Rohdie (Rohdie, 1969) has pointed out, this simple plot situation allows a whole series of oppositions to be set up. To illustrate the shifting valencies of the spatial categories referred to above, the master opposition is not, as in *Sunrise*, country versus city, but small town versus metropolis. Like the shifting master antinomies, the sub-oppositions too begin to shift and turn in the course of the film:

metropolis	*small town*
words	deeds
hypocrisy	honesty
cash relations	human relations
opera	brass band
high art	popular art
individual	community
culture	nature
wit	fun
excess	modesty
sane (mad)	mad (sane)

It was not accidental that the inter-war period in America should throw up films such as *Sunrise* and *Mr Deeds Goes to Town*. They represent America talking to itself about the great transition to modernity which, like everything else in that society, was severely telescoped. If Oscar Wilde's observation that the United States was the only society to have passed from barbarism to decadence without the intervening stage of civilisation overstates the case, the following statistics indicate the rapidity of the transition:

> In 1810 over 90 per cent of the US population was classified as rural, many were self-sufficient farmers. Even as late as 1880 the farm population was 44 per cent of the total population. One hundred years later this figure had dropped to under 3 per cent.
>
> (Short, 1991: 104)

It was in the inter-war period, however, that the United States – demographically and in terms of organised political power – could truly be said to have become an urban society and rural America, as expressed in Prohibition, Biblical fundamentalism and nativist organisations such as the Ku Klux Klan, went down fighting tooth and nail, licking its wounds until the moment, the 1980s and beyond, when it would reconstitute its alliances and attempt, once more, to remake America in its own image. That the so-called 'moral majority' was able to do this with some degree of success may be partly explained by the survival and enhancement of anti-urban, pro-small town/agrarian/pastoral/wilderness ideologies in American life, even (perhaps especially) among city dwellers. The valorisation of the small town and its values is a consistent theme in the films of Frank Capra, the director of *Mr Deeds Goes to Town*, even if, as in *It's a Wonderful Life* (1946), Capra's defence has to invoke desperate fantasies. A similar celebration of the small town is at work in the Andy Hardy films (a cycle starring Mickey Rooney which ran from the mid-1930s to the mid-1950s) and in such films as *Meet Me in St Louis* (1944). As has been stressed above, all discursive spaces have volatile valencies with the same space being deployed to signify quite incompatible ideological positions. This is no less true of the small town in Hollywood cinema (Roffman and Simpson, 1994) which, as well as signifying all that is supposedly good in America, has also been made to stand as an icon of bigotry, small-mindedness and explosive violence, as in *Fury* (1936), *Intruder in the Dust* (1949), *Bad Day at Black Rock* (1955), *Easy Rider* (1969) and, deliriously, *Blue Velvet* (1986). In fact, the two traditions of representation continue to exist side by side and despite (or perhaps because of) increasing demographic evidence of the growth of cities, the small town continues to function in American cinema as a balm to hurt minds and bodies. This is strikingly so in as recent a film as *Sleeping with the Enemy* (1991). As is so often the case, the small town here operates as the binary antinomy not, strictly speaking, of the city itself but of 'city values'. The Julia Roberts figure, physically and emotionally abused by her Wall Street commodities broker husband (Patrick Bergin), fakes her own drowning and moves from their chilly, modernist Cape Cod house to a small midwest town where she forms a relationship with a university drama teacher (Kevin Anderson). Interestingly, the bleak Cape Cod beaches are, with Bergin's New York skyscraper office, the structural opposite of the warm, homely midwest town. In other discourses, of course, the sea coast could be made to signify positive qualities in opposition to the city, as in *Local Hero* (1983). As with *Sunrise* and *Mr Deeds Goes to Town*, the master antinomy of city/Cape Cod versus midwest town is supported by innumerable sub-antinomies as follows:

city/Cape Cod	*midwest town*
concrete/steel/glass	wood
commodities broker	humanities teacher
haute cuisine	apple pie
haute couture	'Laura Ashley' dress
sand	garden
Berlioz	Van Morrison
moustache	beard
suit	jeans
etc.	*etc.*

When Julia Roberts first arrives by bus in the midwest town, her face shows palpable joy as, from her viewpoint, the camera picks up the townspeople going about their everyday activities. The entry of individual characters to a new milieu – in the case of the Julia Roberts figure from (putative) city to small town, but often a rural figure entering a city – is frequently when oppositional ideologies are most sharply posed, as in the entry of Clint Eastwood's western sheriff to New York in *Coogan's Bluff* (1968); Jon Voight's cowboy hustler entering the same city in *Midnight Cowboy* (1969) and Kirk Douglas' wandering cowboy entering Duke City in *Lonely Are the Brave* (1962). There is a similar entrance to New York, this time by ferry across the Hudson, in a film much touted as a significant statement about the city and modernity, King Vidor's *The Crowd* (1928). In fact, *The Crowd* sends out conflicting signals about the city. On the central protagonist's entry to the city, a montage of urban scenes – milling crowds, bumper to bumper automobiles and aerial views of houses packed together – culminates in a series of vertiginous shots up the face of skyscrapers and into a massive open-plan insurance office where hundreds of workers, including the central protagonist (with his own assigned number), sit in serried ranks of identical desks. Alongside this powerful statement of the city's impersonality, the urban milieu is represented as a site of romance, excitement and enjoyment, as in the ride through the New York streets in an open-topped bus and the visit to Coney Island. Though there are repeated references to 'the crowd' in the intertitles (the film is silent), the film does not sustain the expressionist representation of the introduction to the city and the scenes away from the city – the opening in a small town, the honeymoon at Niagara Falls and the family picnic at the seaside – are not constructed as ideological opposites to the city as in *Sunrise*. Niagara Falls, for example, functions as an image of sexual passion, as it was to do later, more feverishly, in *Niagara* (1953), starring Marilyn Monroe. The driving ideology of *The Crowd* is that of Horatio Alger – making it and getting a lucky break.

Coincidentally, two figures associated with the representation of the city – the actor Gary Cooper from *Mr Deeds Goes to Town* and the director King Vidor from *The Crowd* – came together in 1949 to make a film which cannot be ignored in any discussion of the cinematic city – *The Fountainhead* (Plate 1.1) Adapted by Ayn Rand from her own novel, *The Fountainhead* is about the struggles of an architect (based it is said on Frank Lloyd Wright) to have his work accepted in the teeth of philistine opposition. Locatable in a cycle of anti-communist films which Hollywood produced in the late 1940s and early 1950s, *The Fountainhead* is a product of the wilder shores of the American right. Although it fitted perfectly into the hysteria and paranoia of the Cold War, the novel itself had been drafted in the late 1930s when its particular target had been F. D. Roosevelt's New Deal. *The Fountainhead*'s cinematically breathtaking, but ultimately demented, *mélange* of anti-intellectualism, anti-collectivism, sado-masochism, barely repressed homo-eroticism, architectural gigantism and pathological individualism makes the description 'crypto-fascist' not unreasonable, if the phrase is understood as indicating a psychological disposition rather than an organised political movement.

1.1   *The Fountainhead*

The image of the city offered in this *triomphe de la ville* is of a series of towering, Corbuserian skyscrapers produced as an act of will by their creator and existing in no relationship whatever with the society around them. *The Fountainhead* demonstrates that, as with Italian Futurism, modernism is as easily mobilised by the right as by the left. The reference to Italian Futurism is not entirely fortuitous. Although certain of the images in *The Fountainhead* fleetingly suggest the films of Leni Riefenstahl, the image at the end of the film of the architect (Gary Cooper), clad in black, hands on hips and legs apart, standing atop the highest skyscraper in New York (which, of course, he has built), is particularly evocative of the fascist *squadrone* as represented in inter-war Italian graphics, including some by Futurists.

However, *The Fountainhead* is also aesthetically and politically explicable within the trajectory of American modernism. One of the New Deal's central maxims, 'Getting America back to work', extended into the area of art production. The particular mechanism that put *artists* back to work was the Works Progress Administration which, by 1936, was employing over 6,000 artists, each drawing

$28.32 a week for producing work commissioned by the government. The particularly favoured form was mural paintings in post offices and other public buildings. Often the murals depicted significant episodes in local or national history and were executed in a broadly realist (and, therefore, implicitly anti-modernist) style. It was quite logical, therefore, for Ayn Rand, in attacking New Dealism, also to attack its 'official' aesthetic and to valorise its opposite, High Modernism. Rand's endorsement of modernism from the extreme right was supported by changes in the American art world taking their impulse from quite another political quarter. Throughout the 1940s an important turn towards High Modernism was taking place within American art. Indeed, it has been said that, at this time, the centre of gravity of modern art moved from Paris to New York (Guilbaut, 1983). This was to come to full flowering, at the level of art practice, in the 1950s movement of Abstract Expressionism in which the central figures were Jackson Pollock, Willem de Kooning and Mark Rothko. The ideological underpinning of this movement was provided principally by two critics, Clement Greenberg and Harold Rosenberg. Greenberg, in particular, had been active on the left in the 1930s but seems to have become disillusioned after disclosures such as the Moscow treason trials. It would seem that, in the field of American art, anti-Stalinism mutated into Trotskyism and then into 'art for art's sake'. Although figures such as Greenberg and Rosenberg retained something of the rhetoric of the 1930s' left, their writings increasingly stressed that the act of artistic creation was for its own sake and that of the creator, sentiments recurrently voiced by Howard Roark, the architect in *The Fountainhead*. Thus it was, in America in the late 1940s, through a complex set of overdeterminations, that art production became prised away from social meaning and defined primarily as self-expression. As such it became mobilisable within American Cold War rhetoric, achieving its most unhinged statement in *The Fountainhead*. However, it has been pointed out (Petley, 1988; Walker, 1993) that there is a profound contradiction at the heart of *The Fountainhead* both as novel and as film. Trumpeting the virtues of High Modernism and sneering at the art of the past, *The Fountainhead*, in its literary and cinematic forms, displays all the features of the classic narrative structure of the Victorian novel.

Such was the overwhelming psychological effect of modernity, and within it the oppressiveness of the city, that even film-makers relatively unconcerned with the milieux in which their films were set, were affected by it.

Alfred Hitchcock, in the opening sequences of *Shadow of a Doubt* (1943), offers an intriguing glimpse of what might have been had he become preoccupied with the transition to modernity rather than simply living within it. Very much like the opening of the same director's *Psycho* (1960), *Shadow of a Doubt* begins with a series of increasingly narrowing cityscapes until the film enters the room of the central protagonist, Uncle Charlie (Joseph Cotton). The rundown nature of Hitchcock's city – foreshadowing the image of the city as urban wasteland so central to the American cinema of the 1980s and 1990s – is rendered even more strange by the oblique angle of some of the camera setups; the strange music on the soundtrack, including an off-key version of the *Merry Widow* waltz; and the fact that the police are closing in on Uncle Charlie, a modern Bluebeard who disposes of rich widows. The bleak cityscape is immediately succeeded by the place to which Uncle Charlie goes to lie low for a while, Santa Rosa, California, the archetypal American small town with, instead of detectives closing in for the kill, a genial middle-aged traffic cop and, instead of the sinister, off-key strains of Franz Lehar, the kind of soundtrack music that signifies 'normal, placid, small-town America'. In some respects the introduction to Santa Rosa in *Shadow of a Doubt* foreshadows Julia Roberts' entry to the midwest town in *Sleeping With the Enemy*. However, having set up the city/small town opposition, Hitchcock does not follow it through but moves swiftly to his own recurrent obsessions: the transference of identity and the transference of guilt that goes with the secret knowledge of murder having been committed.

Film critics are generally agreed that the gangster film and its generic affiliates such as the *film noir* and the *film policier* occupy a special place in the representation of the city. This insight is at the core of the first serious critical account of the genre:

> The gangster is the man of the city, with the city's language and knowledge, with its queer and dishonest skills and its terrible daring, carrying his life in his hands like a placard, like a club . . . for the gangster there is only the city; he must inhabit it in order to personify it: not the real city, but the dangerous and sad city of the imagination which is so much more important, which is the modern world.
>
> (Warshow, 1962: 131)

A later critic has widened the argument beyond the gangster film and its generic affiliates to suggest that the cinema was the structure-in-dominance among the determinations which produced a particular American social type, the City Boy:

> The City Boy was a product of performance, genre and ideology transmuted into popular entertainment. He did not so much mirror social life, if contemporary observers are to be believed, as create a model for life to imitate him. Arising from the teeming ethnic polyglot of the modern industrial city – especially New York – he began playing a central social role with the more or less simultaneous occurrence, in the late 1920s, of talking pictures and the Great Depression. His rise to prominence alongside the cowboy as a major figure in the representation of American manhood seemed to suggest important changes in concepts of male behavior – of individualism in relation to social constraint, of sexuality, romance and family life.
>
> (Sklar, 1992: xii)

It is no accident that the particular 'city boys' whom Sklar discusses, James Cagney, Humphrey Bogart and John Garfield, were key players in the gangster film and its generic affiliates.

Although the gangster film's antecedents have been traced to times well before 1930 (Vernet, 1993), it was in that year that the film appeared which is generally conceded to be the first fully recognisable gangster film – *Little Caesar* (Plate 1.2). Making explicit what was to remain implicit in most other gangster films, *Little Caesar* dramatises the tension between the old, agrarian European societies from which the gangsters sprang and the heartless, wide-open world of the American city in which they make their way. This tension is particularly acute in one gang member, on the one hand seduced by the money and the easy life of the city and, on the other, drawn to his Italian-speaking mother, his home and his church. It is this figure who, in a scene which was to become archetypal of the genre, is shot dead in the street from a moving car as he mounts the steps of the church to confess his involvement with crime.

As many critics (McArthur, 1972; Shadoian, 1977; Rosow, 1978; Hirsch, 1981) have pointed out, so central is the city as a presence in the gangster film, the *film noir* and the *film policier* that it and its milieux very often figure in the titles of films: *Big City Shadows* (1932); *Scarlet Street* (1945); *The Street With No Name* (1948); *Cry of the City* (1948); *The Naked City* (1948); *City Across the River* (1949); *Dark City* (1950); *Night and the City* (1950); *Where the Sidewalk Ends* (1950); *The Asphalt Jungle* (1950); *Down Three Dark Streets* (1954); *The Naked Street* (1955); *While the City Sleeps* (1956); and so on. Aside from a sub-genre of films about (often) midwestern Depression criminals, from *High Sierra* (1941) to *Bonnie and Clyde* (1967), the action of the gangster film is played out within circumscribed urban milieux: dark streets; dingy rooming-houses and office blocks; bars and nightclubs; precinct stations; and luxury penthouses. As mentioned above, a recurrent motif of the form is the gangster being shot down in the street or, memorably in *Underworld USA* (1961), being shot elsewhere and crawling into an alley to die.

One of postmodernism's most self-congratulatory features is the evaporation of the traditional distinction between High and Mass Art.[3] Clearly, this has had a beneficial effect to the extent that any act of discourse production, no matter how 'lowly', is grist to the mill of serious analysis. However, there are two less useful outcomes, a refusal to recognise as pertinent the substantial rhetorical differences between High Art and Mass Art, and a purblindness about the extent to which the two areas have been historically interpenetrated, not least in Hollywood cinema. One of the great Modernist (and therefore High Art) responses to the city was Paul Citroën's photomontage *Metropolis* (1923) which conveys a vertiginous sense of 'cityness', capturing both the sense of the city's oppressiveness and its breath-taking excitement as felt by part, at least, of the population of Weimar Germany. While it is often remembered that an analogous form of cinematic montage was practised, and extensively theorised, in the Soviet Union in the 1920s, it is usually overlooked that, from the 1930s, Hollywood cinema practised its own form of montage, doubtlessly influenced by what was going on in the Soviet cinema, and that the American practice posed no problems of readability for a mass audience. The characteristic Hollywood montage sequence is very short and comprises several

1.2   *Little Caesar*

visual and sound images superimposed upon each other by way of dissolves. Often signifying the passage of time and space, such a sequence would frequently deploy images of fluttering calendar pages, train wheels, place names and newspaper headlines. While the gangster film was by no means the only site for such material, the outbreak of gang warfare and its extension over time, or the gang's reign of terror extended over space, would often be signified by a montage sequence. What is important from the point of view of representations of the city is that such a moment in a gangster film – the one in *G-Men* (1936) is a good example – usually made up of speeding automobiles, blasting machine-guns and newspaper headlines, offers a condensation of the idea of 'cityness'. Indeed, a freeze-frame from such a sequence, taken from the heart of a piece of Mass Art, would not be dissimilar to certain High Art photomontages of the same period.

If, after *Little Caesar*, the country/city antinomy tended to recede and the city remain as the expressionist milieu which shaped the characters' lives, it was to re-emerge from time to time, as in *On Dangerous Ground* (1952) and *Witness* (1985). In both films city police officers are forced to go into the country, but the two films' handling of the country/city opposition is very different. In *Witness*, while violent acts occur in the city, it is not represented as inherently evil, but the country – in the shape of an agrarian Amish community – is constructed as sentimentally arcadian. A much more profound film than *Witness*, *On Dangerous Ground*'s explosively violent policeman (Robert Ryan) is empty and alienated *because* of the city and what his job there requires him to do. His journey to the country becomes a spiritual quest which ends in a kind of redemption. The country in *On Dangerous Ground* is no prelapsarian Eden. It is a pitiless midwinter wilderness inhabited by figures capable of ferocious acts of violence, but the policeman is nevertheless humanised and transformed by confronting the violence within himself. The philosophical (and, indeed, theological) underpinning of *On Dangerous Ground* becomes explicit from time to time. A fugitive boy is represented in terms of animal imagery and the boy's blind sister, with whom the policeman forms a tenuous relationship, is called Mary Walden, a name which invokes Henry David Thoreau and locates the film right at the heart of American debates about past and present, ruralism and urbanism, agrarianism and industrialism, and individuality and community.

As might be expected, the worsening condition of 'real' inner cities in the last two or three decades has been paralleled by cinematic representations of the city as a desolate battleground traversed by human monsters on the very margins of sanity. Such a view is signalled in films such as *Death Wish* (1974), *Taxi Driver* (1976) and *Hard Core* (1979) and has been accurately described by Frederic Jameson as 'a new Third World space within the First World city' (Jameson, 1994). This dystopian city has been realised most potently in certain of the films of John Carpenter, particularly *Assault on Precinct 13* (1976), *Escape From New York* (1981) and *Big Trouble in Little China* (1986). In *Escape From New York* the dystopian view of the city is taken to its imaginative extreme when, set in a period only fifteen years ahead of the date the film was made, Manhattan has become a hermetically sealed penal colony for America's most violent prison inmates. It is not coincidental that Carpenter is also the co-translator of *Report From the Besieged City and Other Poems* (Herbert, 1985). In some respects associated with the Carpenterian view of the city as desolate battleground, but in other respects dissimilar, are two other perspectives on the city which might be called the comic book view and the postmodern view. Both further loosen the hinges which have held cinematic representations of the city in some contact with 'real' cities. The comic book view is exemplified in *Batman* (1989) and *Dick Tracy* (1990). Signalling the complete collapse of time and space in cinematic representation, the postmodern city has been realised most forcefully in the work of Ridley Scott, its fullest expression generally conceded to be *Blade Runner* (1982). The shrill, postmodern city of *Blade Runner* – all concrete, glass and neon and its cosmopolitan denizens indistinguishable from the manufactured replicants who move among them – bears some resemblance to the actual Japanese city of Osaka which is the main setting of Scott's 1989 film, *Black Rain*. Peter Wollen conflates the comic book city and the postmodern city, seeing both, with some justice, as depicting 'the post-Fordist city of deindustrialization, casual and freelance employment, large-scale immigration, the privatization of welfare, and social polarisation' (Wollen, 1992: 26).

The hegemonic ideologies of American life valorise the country over the city. Where, then, in American cinema are sympathetic representations of the city to be found?[4] When, in a 1930s gangster movie, a city criminal holed up in the country says, 'I don't like the country. The crickets make me noivous' and when, in *Sweet Smell of Success* (1957), Burt Lancaster's megalomaniac newspaper columnist looks around the New York streets and says, 'I love this dirty old town', the city can hardly be said to have sympathetic mouthpieces. Less reprehensible enthusiasm for the city in the American cinema is discernible principally in two areas. One is the American film musical. Although there are to be found from time to time downbeat and even sinister elements, the Hollywood musical is usually maniacally upbeat. Richard Dyer (Dyer, 1977) has talked about the Hollywood musical's 'utopian sensibility' and has analysed its workings through five categories: energy, abundance, intensity, transparency and community. All of these categories, wherever they are deployed, are liable to produce congenial milieux. The Hollywood musical sometimes creates rural utopias, as in *Oklahama!* (1955) and *Carousel* (1956), and one of the screen's most venomous representations of the city is in *Brigadoon* (1954) where the movement from the faery village of the title to New York is accomplished by one of the great transitions in the history of cinema. The tranquillity of Brigadoon gives way to a vertiginous view of the New York skyline accompanied, on the soundtrack, by harsh, discordant, 'modernist' brass music. However, the Hollywood musical is also the site of the most affirmative representations of the city. This is evident in the abstractly signified 'cityness' of certain sets in the backstage musicals of *42nd Street* (1933), backdrops for intense expressions of *joie de vivre*, but most triumphantly in *On the Town* (1949). As Dyer (1977: 12) describes it: 'what makes *On the Town* interesting is that its utopia is a well-known modern city. The film starts as an escape – from the confines of navy life into the freedom of New York.' Again and again in the Hollywood musical 'real' cities are transformed into utopian spaces: Rio de Janiero in *Flying Down to Rio* (1933); Paris in *An American in Paris* (1951), *Funny Face* (1957) and *Gigi* (1958). The bleak, rain-soaked streets of the *film noir* become, in a musical like *Singin' in the Rain* (1952), the site for delirious outpourings of happiness and joy in living.

The other major site for an affirmative view of the city is the work of Woody Allen whose films could be described collectively as a continuing love affair with New York.[5] So central is that city to Allen's life and art that one critical work is called simply *Woody Allen: New Yorker* (McCann, 1990) and yet another (Spignesi, 1992) includes a chapter on the urban geography of three of his films. Allen's celebration of New York is at its warmest and most intense in the opening sequence of *Manhattan* (1979). Justly famous for its monochrome beauty, it deploys – to the accompaniment of George Gershwin's *Rhapsody in Blue* – several images of New York including the skyline, the Empire State Building, the Brooklyn Bridge and Central Park. Attempting to find an appropriate narrative voice, the off-screen narrator makes several references to the city: the fact that New York exists for him 'in black and white and pulsated to the great tunes of George Gershwin'; that 'New York meant beautiful women and street-smart guys who seemed to know all the angles'; and that 'New York was his town and always would be'. The narrator of *Manhattan* is here traversing the same terrain as Thaw at the beginning of this essay.

Clearly, then, Hollywood cinema consistently takes 'real' American cities such as New York and San Francisco (as in *Vertigo*, 1958 and *Bullitt*, 1968) – the 'real' being in inverted commas because they are already functioning discursively – and reinscribes them into discourse once more, predominantly those discourses about the quality of the 'natural' and the built world through which meaning has been imposed on the transition to modernity. But, as the above references to *An American in Paris*, *Funny Face* and *Gigi* indicate, it is not only American cities which have been so 'discursified'. Roughly speaking since the end of the First World War, Hollywood has operated both an economic and aesthetic hegemony over world cinema (Ellis, 1982; Thompson, 1985). For instance, of all films shown in British cinemas in 1992, 85.8 per cent were in every sense Hollywood films and a further 4.6 per cent were Hollywood-financed though made in Britain (Leafe and Illott, 1994). Although the figures differ from one national culture to another throughout the world, the broad argument about Hollywood dominance holds good. The ideological outcome of this is that many national cultures, where they are not *wholly* dependent on Hollywood movies for cinematic representations of themselves, are certainly faced with a situation in which Hollywood movies offer the most popular and the *dominant* representations. With regard to the cinematic representation of geographical space, both urban and non-urban, sense of such space is overwhelmingly derived from Hollywood films. The problem is that, since the point of utterance of such representations is American, they reflect an American perception of the world. In short, Hollywood has created a series of Others which in no sense relate to the self-definition of these diverse other places and peoples: rather they project the needs, fears, fantasies and representations of particular American ideologies. It is necessary to de-individualise this question. It cannot be said that Hollywood movies simply promote the view of the world of those who make the movies. For instance, despite the substantial presence of people of Jewish origin in powerful positions in Hollywood, the ethnic and ideological norms in Hollywood movies have consistently been WASP, although this may be changing (Friedman, 1991). With regard solely to the representation of cities, there must hardly be a major city in the world which (to inflect Thaw's argument at the beginning of this essay) is not known primarily by way of Hollywood. Like Warshow's gangster city, these too are cities of the imagination, heavily imbricated with America's sense of the Other and characteristically introduced by a familiar landmark (e.g. the Eiffel Tower, the Coliseum, the Brandenburg Gate, Big Ben) and a few bars of 'ethnically' significatory music. Thus, for example, *An American in Paris* constructs that city primarily as the centre of world art and as a fountain of inspiration for American artists; *Funny Face* sees Paris as the locus of a comic, and probably bogus, intellectualism; *Three Coins in the Fountain* (1954) and *The Roman Spring of Mrs Stone* (1961) suggest that Rome is primarily a source of romance between American tourists and impecunious Italian aristocrats; the Berlin of *Cabaret* (1972) is shaped within the dominant narrative of Weimarian decadence (the point of view of National Socialism, incidentally); and the Moscow of *Telefon* (1977), suffused with Cold War ideology, is dark, cold and menacing.

While taking note of Hollywood's post-First World War hegemony in the cinematic representation of world space, it should not be suggested that Hollywood is the point of origin of all of the discourses within which spaces and places are constructed. The cinematic representation of London is instructive in this regard. There is, in the film *Ziegfeld Follies* (1946), a dance number involving Fred Astaire and Lucille Bremer and entitled 'Limehouse Blues'. Set in an indeterminate, pre-Second World War London, it shows a bleak, fog-ridden Thames-scape populated primarily by Chinese but also by stalwart 'bobbies', Bill Sykes-type ruffians, pawky and cheerful pearly kings and queens, and bizarre down-and-outs. Where does this representation of London, in a Hollywood movie of the 1940s, come from? There is a specific cinematic reference point, D.W. Griffith's *Broken Blossoms* (1919), but this simply pushes the question back by a generation. It is quite likely that the 'London discourse' informing both *Broken Blossoms* and the 'Limehouse Blues' section of *Ziegfeld Follies* is a composite discourse deriving from travellers' accounts of London, the novels of Charles Dickens (particularly *Bleak House*), the Sherlock Holmes stories of Sir Arthur Conan Doyle, and press accounts of the Jack the Ripper murders. It is the same discourse (or, more accurately, part of the *bricolage* making it up) which underpins the American popular song *A Foggy Day in London Town*. Certainly, this construction of London (or diverse aspects of it) is evident in many Hollywood-based or Hollywood-financed movies such as *Dark Eyes of London* (1937), *Gaslight* (1944), *Hangover Square* (1945), *The Verdict* (1946), *Night and the City* (1950) and, in a more upbeat form, *Mary Poppins* (1964). Although the London of *Ziegfeld Follies* is in colour, the colour (apart from Lucille Bremer's yellow dress) is very muted. The 'London discourse' shaping the above films may have found the harsh black and white of the classic *film noir* a more congenial form of expression, hence the clustering of these titles around the 1940s.

It is never a question of a discursive view of geographical space giving way to a more 'realistic' view. There is only the possibility of other discourses arising to compete with existing ones. Put another way, realism is itself a discourse, a convention of representation which might perhaps be better described as 'the realist effect'. Thus, there have been several 'London discourses'. Offering a more detailed discussion of the cinematic representation of London than is possible here, Sutcliffe (1984) suggests that a new cinematic discourse relating to London emerged through the work of those figures from the British Documentary movement of the 1930s who were drafted into making propaganda films (always a feature of that movement) about the Second World War. The orientation of this new 'London discourse' is summed up in the title of one such film, *London Can Take It* (1940), which mobilises particular London landmarks such as the dome of St Paul's and Vaughan Williams' *London Symphony* to create a narrative about a proud city enduring under the bombardment of the Luftwaffe. This more upbeat discourse mutated in the post-war period into active 'boosterism' of place (Gold and Ward, 1994) as the wartime documentary movement opened out into the making of more diverse, sometimes commercially sponsored films dealing, among other things, with post-war planning and tourism. However, to talk about a specifically urban discourse is to extrapolate from the 'discursive bundle' which makes up any particular film. To do full justice to the complexity of *London Can Take It*, it would have to be related to that curious melding of Surrealism and Englishness which informs the entire *oeuvre* of the film's maker, Humphrey Jennings (Petley, 1978). Nevertheless the less fog-ridden London images of the wartime documentaries, which carried over into post-war London-based feature films such as *Waterloo Road* (1945) and *The Blue Lamp* (1950), marked a development in the cinematic representation of London as did the all-pervading discourse of 'swinging London' – traceable across journalism, advertising, fashion, gossip and anecdote – which arose in the 1960s. Energetic, colourful, demotic and, ultimately, totally superficial, it is perhaps best signalled in the Italian Michelangelo Antonioni's 1966 film *Blow-Up*. If, however, as Homi Bhabha (1990) suggests, 'the historical and cultural experience of the western metropolis *cannot* now be fictionalized without the marginal, oblique gaze of its postcolonial migrant populations cutting across the imaginative metropolitan geography of territory and community, tradition and culture' (cited in Donald, 1992: 455), then this fact will increasingly register its presence in cinematic representations of London. It has already done so most notably in *My Beautiful Laundrette* (1985) while at the same time signalling the presence in British culture (albeit in a less deranged form) of the dystopian view of the city as urban battlefield so markedly present in contemporary American cinema. Peter Wollen reads films such as *Sammy and Rosie Get Laid* (1987), *The Last of England* (1987), *The Cook, the Thief, His Wife and Her Lover* (1989) and the London sequences of *Brazil* (1985) as lucidly anti-Thatcherite and, as such, connecting with the economic, social and political forces shaping the cinematic representation of American cities as urban wastelands.

Although Hollywood's economic, psychological and aesthetic hegemony requires that its negotiating of the great historical transition to modernity be privileged in any discussion of cinematic representations of the city, the same transition was being negotiated in other national cinemas, initially those of Europe but eventually, as a result of uneven economic development, of Third World countries as well. Much recent work has been preoccupied with the changes in perception which have attended the transition to modernity. It has been suggested that

> a series of sweeping changes in technology and culture created distinctive new modes of thinking about and experiencing time and space. Technological innovations including the telephone, wireless telegraph, x-ray, cinema, bicycle, automobile and airplane established the material foundation for this reorientation; independent developments such as the stream of consciousness novel, psychoanalysis, Cubism and the theory of relativity shaped consciousness directly.
>
> (Kern, 1983, cited in Natter, 1994: 224–5)

David Harvey's phrase 'space–time compression' (Harvey, 1989) elegantly subsumes both the material and psycho-aesthetic dimensions of the process which was nowhere more self-consciously confronted with regard to the city and cinema than in Weimar Germany. The cluster modernity/city/cinema was addressed theoretically by Walter Benjamin (Benjamin, 1969) and Siegfried Kracauer (Kracauer, 1947) and practically by a whole range of German film-makers of diverse modernist orientations. Just as the massive presence of the city forced itself into the very titles of films in the American cinema, so too in Weimar Germany was there a series of 'street films' such as *The Joyless Street* (1925), *Tragedy of a Street* (1927) and *The Street* (1929). The psychological weight of the city was given monstrously expressionist form in Fritz Lang's *Metropolis* (1926; Plate 1.3), but perhaps the most interesting from the point of view of signifying the new sense of urban space is *Berlin, Symphony of a City* (1927; Plate 1.4). Walter Ruttmann's film is abstract and modernist in several ways: its lack of formal plot; its lack of interest in the human beings who inhabit the city; its construction on the basis of musical theory; and its linkages through the formal characteristics of the images. However, its reception in Weimar Germany indicates the extent to which – despite its virtually complete repression of the structural other of the country – it was perceived as an intervention in that very debate about the country and the city which had so informed American cinema:

> As *the* metropolis of Weimar Germany, Berlin was the locus of both desire and anxiety in reaction to modernization much commented upon by contemporaneous social theorists of the left and right. One should not be so awestruck today by the cultural brilliance of the avant-garde, for whom the cosmopolitan

1.3 *Metropolis*
© F.-W.-Murnau-Stiftung, Wiesbaden/Transit-Film. GmbH, Munich. Courtesy of BFI Stills, Posters and Designs.

atmosphere of Berlin represented the best impulses of Weimar Germany, to forget that for a broad spectrum of anti-modernist and folkish Germans, Berlin and all that it stood for was the devil incarnate. Berlin became a crystallization point of resentment against industrialization, capitalism, democracy and the cultural influence of the West following Germany's defeat in World War I. Anti-modernists penned the term 'asphalt culture' to refer to the lack of genuine culture and social values promoted by urban life. The term connoted a loss of direct contact with the soil and the ethical life, an agrarianist ideology attributed to it, as well as the rootlessness (artificiality) of urban life. During the 1920s, this position was also explicitly tied to a critique directed against the democratic government of the Republic and its 'soulless' culture, nowhere more materially manifest than in Berlin. Wilhelm Stapel, a folkish writer and editor of *Deutsches Volkstum*, phrased the debate being waged in Weimar Germany as one between the German landscape and the city. 'The spirit (*Geist*) of the German Folk rebels against the spirit of Berlin. Today's battle cry must be "The resistance of the landscape against Berlin".'

(Natter, 1994: 214–5)

1.4   *Berlin, Symphony of a City*
Courtesy of BFI Stills, Posters and Designs.

Berlin, Symphony of a City was mobilised by both sides of this debate, as an affirmation of the diversity and excitement of modernity and the city, and as a reflection of the city as Sodom. What is so interesting about the above quotation is its revelation that the same discourses about country and city are condensed, in different cultures, into specific, national ideologies which circulate within or are mobilised in response to particular films. Thus, in the United States, discussion of *Mr Deeds Goes to Town* is usually 'cashed' in terms of agrarian populism, while in Weimar Germany such discussion might be 'cashed' in terms of proto-Nazism. Without suggesting that the work of F. W. Murnau displays any such tendency, it is perhaps worth noting that the maker of *Sunrise* – that arch-condemnation of the city – had a distinguished career in Weimar Germany before going to the United States.

In some respects, *Berlin, Symphony of a City* can be seen as the archetype of 'city symphonies' and other modernist cinematic responses to modernity which were appearing in many societies in the inter-war period. Other notable examples are *Manhatta* (1921) in the United States, *Rien Que Les Heures* (1926) in France and *Man With a Movie Camera* (1929) in the Soviet Union. There was even, as Giuliana Bruno has demonstrated (Bruno, 1993), for almost the whole of the first three decades of the century, a Neapolitan city cinema which, unusually, dealt primarily with the lives of women. The phenomenon reached as far afield as Japan which, very much under the influence of German film culture, had its own 'street films' and, in 1929, a 'city symphony'. The impulse found its way even into the interstices of mainline film culture and the 'tenement symphony' in the 1946 Marx Brothers film *The Big Store* may represent an attenuated instance.

The great historical debate about the relative virtues of the country and city would be replayed in the films of many European and non-European societies in the face of the inexorable drift of peasants into cities throughout the world. Luchino Visconti's *Rocco and His Brothers* (1960) begins with the death of a peasant farmer in Southern Italy and the movement of his family to the northern industrial city of Milan. Much of the strange power of Luis Buñuel's *Los Olvidados* (1950), set in Mexico City, comes from the clash between the bizarre rituals and beliefs of the incoming peasantry and those of the more cynical, materialistic sub-proletariat of the city. The tension is apparent even in British cinema in a film such as *Floodtide* (1949) which, through its celebration of shipbuilding on Clydeside, is heavily imbricated with the country/city opposition. The film begins in the country with the explicit rejection of farming by the central protagonist in order to pursue a career in shipbuilding in Glasgow. It has been a recurrent theme of this essay that the dualisms within discourse are fundamentally unstable and capable of being mobilised into incompatible ideologies. Such is the case in *Floodtide* in which the River Clyde is made to function in two contradictory discourses relating to Scotland: the Scotland of dynamic, industrial activity and the Scotland of beautiful hills, lochs and rivers (McArthur, 1982). In the nearly contemporary *The Gorbals Story* (1950) the Clyde's ideological meaning shifts yet again to become the dividing line between working class and middle class Glasgow (Hill, 1982).

As evidence of the extent to which modernisation and urbanisation are increasingly provoking psycho-aesthetic responses throughout the Third World, the Taiwanese film *Homecoming* (1984) deals with two friends separated in childhood, one who has gone from mainland China to Hong Kong and become a publisher, the other who has remained in a mainland village and become a schoolteacher. The former is subject to the problems characteristic of the city dweller throughout the world – fragile economic status, disintegrating family and shattered relationships. It is through her reconnection to a mythic, agrarian China that she is able to come to terms with the drawbacks of the materially oriented city life of Hong Kong. *Homecoming* is quite subtle in its negotiation of the country/city opposition and it is, of course, filtered through specific Chinese ideologies (for example, those relating to family and ancestors) and complicated by the fact that the People's Republic of China has, in the post-Mao period, embarked upon a specific policy of modernisation, an inhibition on unambiguous endorsement of the country side of the country/city opposition. Such a policy has no bearing on those Asiatic societies which have no relationship with mainland China. The South Korean film *First Son* (1985) has been described as '[allegorizing] the rural–urban opposition in stark, moralistic terms' (Wilson, 1994).

Although, as has been argued throughout this essay, there are few more useful structures than the historically far-reaching country/city opposition for understanding the way in which cities have been represented on film, it has to be recognised that this strategy proposes an extra-cinematic determination of the process. In the last analysis, this is entirely proper because films do appropriate and recycle discourses at large in the world outside cinema and ideologies current in specific societies. That said, there is much truth in the view that art is made from other art rather than from 'reality' and, as many a film-maker will testify, the impulse to make a particular film may come as much from exposure to other films as to the events happening around the film-maker in question. In talking about the representation of cities in cinema, therefore, one has to confront the inflection given to such representations, in the post-Second World War period, by the massive international influence of Italian Neo-Realism on films made at this time. Neo-Realism, like any other artistic movement, incorporated the most diverse temperaments and practices – at one extreme the austere dedramatisation of Roberto Rossellini, as in *Paisa* (1946), at the other the sonorous melodrama of Vittorio de Sica, as in *Bicycle Thieves* (1949) – but some of the elements that these films dubbed Neo-Realist had in common were location rather than studio shooting, the use of non-actors, and the tracing of the impact of social (and sometimes metaphysical) forces on the weakest members of society (e.g. women, children, unemployed workers). Such films were seen and greatly admired by critics and cineastes throughout the world and their influence is discernible in every national cinema in the decade or so following the Second World War. To illustrate the critical problem in question, a recurrent image in three post-war films from diverse societies is of children moving through the ruins of war-torn cities. However, underneath the identical visual look of these films quite different things are going on. Aleksander Ford's *Five Boys From Barska Street* (1952) is concerned with the implacable working of social and historical forces on the lives of the five boys; Roberto Rossellini's *Germany, Year Zero* (1947) is more interested in the corruption of childhood by the Nazi legacy; and Charles Crichton's *Hue and Cry* (1946) is simply an engaging English comedy. Clearly, despite the diversity of their ideological projects, the manifest influence of Italian Neo-Realism on their overall look, and the foregrounding of children as protagonists, all of these films offer images of the war-torn fabric of their respective urban milieux and, as such, create new images of Warsaw, Berlin and London.

When people respond with recognition and pleasure to cinematic representations of the spaces they inhabit in the 'real' world, this should not be dismissed with academic arguments about our incapacity to know the world except through discourse. At the same time, the act of filming any space is clearly an act of discourse production (see note 4), prey to complex elision, condensation and repression and as dependent on previous acts of discourse production as on relationships with the 'real' world.

This is a tension we must simply live with.

NOTES

1   In a classic case of throwing the baby out with the bath water, some post-modernisms – seeking to deconstruct the binary oppositions so dear to structuralism on account of (among other things) their alleged privileging of one half of the dualism (e.g. Man/Woman, Nature/Culture) – have abandoned binarism entirely. This is to discard one of the most useful critical tools for dealing with (particularly popular) culture. The strategy of this essay is to retain binarism but to insist on the instability of meaning on both sides of any dualism, on their capacity to signify even incompatible positions in different discursive situations. For example, 'small town' in *Mr Deeds Goes to Town* and *Sleeping With the Enemy* signifies quite the reverse of what it signifies in *Easy Rider* and *Blue Velvet* and 'city' likewise in *Sunrise* and *On the Town*. To reassert classical structuralist binarism, in each case the meaning of 'small town' or 'city' derives from what it has been set against in the particular discursive context.

2   The central emphasis in this essay is on Hollywood cinema and the cinema of Weimar Germany. That this is so is due to the forcefulness of representations of the city in these cinemas. In a pleasing homology, very likely connected with this fact, the history of discourse about the city has been categorised (Sennett, 1969) into a German school, represented primarily by figures such as Max Weber, Georg Simmel and Oswald Spengler and an American (Chicago) school represented by Robert Park, Louis Wirth and Ernest Burgess. There is some overlap between the two schools, possibly due to the fact that Robert Park had been Simmel's student at

the University of Heidelberg before the First World War. What the two schools had in common was a perception of the effect of cities on individuals in the sense of fragmenting lives into different compartments and relationships as opposed to the seamless continuity of pre-urban living. It is not academic works themselves which enter general consciousness, but rather journalistic commentaries and popular appropriations of them. At the risk of considerable oversimplification, it might be said that Weimar Germany tended to appropriate the Spenglerian view of the city as oppressive and alienating while America appropriated the ideas of Robert Park, particularly the suggestion that the city offered the promise of human growth and the possibility of non-conformity. Such emphases were to some extent incorporated into these societies' cinematic representation of cities. For instance, it is tempting to see the popularisation of the Chicago school's ideas underlying that cycle within the gangster film which, in the mid-1930s, offered a social theory of crime. The best-known films of that cycle were *Dead End* (1937) and *Angels With Dirty Faces* (1938), both of which register the significant presence of the city.

3   The history of the High Art/Mass Art opposition, its centrality in certain Hollywood movies – *Shall We Dance?* (1937), *Sullivan's Travels* (1941), *Do You Love Me?* (1946), *Three Daring Daughters* (1948), *The Band Wagon* (1953) – and its eventual collapsing by postmodernism deserves a book in its own right. Such a study would range across all the arts and, as in the example given here of montage *vis-à-vis* Hollywood cinema,

would have to take account of the historical interpenetration of the two modes as in the involvement of classical *virtuosi* with popular musical forms. In many respects an archetypal figure in this postmodernist conflating of modes is the classical violin *virtuoso* Nigel Kennedy. His ear studs, spiky hair, creatively tailored dress suit and loud bow tie, 'estuary' accent, passion for Aston Villa, tendency to introduce his rendering of Vivaldi as 'a bit of Viv', and dropping of the 'communing with the Infinite' demeanour in favour of a laughing photograph on his record sleeves, is anathema to an older generation of classical music lovers. While much of Kennedy's style may be a cynical commercial attempt to 'cross over' and incorporate a further sector of the paying public, there is probably an important shift of sensibility taking place as well.

4   While Note 1 enters a plea for the retention of binarism, it is appropriate here to sound a note of caution. The discussion of cinematic representations of the city has been structured thus far under the broad opposition negative/affirmative. The note of caution is about the extent to which any discourse says *anything* about the 'real' space it signifies as opposed to the moral/ideological frameworks within which the enunciators of the discourse live. This problem is explored at length with regard to the literary representation of geographical space by Marx (1980). However, a distinction may have to be made between literary and photographic/cinematic discourse because of the *iconic*, in the Peircian sense (Peirce, 1977), dimension of the latter, because of the existential connection between what is set before the camera and what appears on the resultant film. A concrete example of this is the evident sense of recognition and pleasure we display when confronted with photographs of ourselves, those we know and the spaces we inhabit. This pleasure exists in tension with another response, the sense of otherness produced by mechanical features of the photographic apparatus and the photochemical features of the film stock, as when the latter picks up and foregrounds a red cushion cover we have hardly noticed in the pro-filmic event (i.e. the situation photographed). When we remark that the thing which stands out most in the photograph is the red cushion cover, we are recognising, implicitly at least, one dimension of photography's complex discursivity.

5   New York's centrality in both affirmative and negative representations of the city is recognised in Barbera *et al.* (1986) which includes a filmography of about 500 films set in New York, from *The Jazz Singer* (1927) to *Hannah and Her Sisters* (1986); an account of the recurrence in these films of particular sites (e.g. Broadway, Central Park, Greenwich Village, the Empire State Building, Harlem); and extracts from interviews with diverse film directors (e.g. Woody Allen, Charles Chaplin, Jules Dassin, Alfred Hitchcock, Sergei Eisenstein, Elia Kazan, Fritz Lang, Roman Polanski and Martin Scorsese) who have made observations about the city. Hillairet *et al.* (1985) offers an analogous account of the representation of Paris in cinema, though mainly in the work of avant-garde film-makers, and Brunetta and Costa (1990) discuss the more general representation of cities in film, although again mainly in the work of modernists such as Ruttmann, Cavalcanti, Dziga Vertov (see also Michelson, 1985), Man Ray, Moholy Nagy and Richter.

Niney (1994) discusses the cinematic representation of 'real' cities such as Paris, Berlin, London, Rome and Prague as well as such cities of the imagination as 'Tativille' – the urban space constructed in the films of Jacques Tati.

REFERENCES

Barbera, A., Cortellazo, S. and Tomasi, D. (eds) (1986) *New York, New York: La Citta, Il Mito, il Cinema*, Torino: AIACE.

Bhabha, H. (1990) 'Novel Metropolis', *New Statesman and Society*, 16 February.

Benjamin, W. (1969) *Illuminations*, New York: Schocken.

Brunetta, G. P. and Costa, A. (eds) (1990) *La Citta Che Sale: Cinema Avanguardie, Immaginario Urbano*, Trento: Manfrini Editori.

Bruno, G. (1993) *Streetwalking on a Ruined Map: Cultural Theory and the City Films of Elvira Notari*, Princeton, NJ: Princeton University Press.

Donald, J. (1992) 'The city as text', in R. Bocock and K. Thompson (eds) *Social and Cultural Forms of Modernity*, London: Polity Press/Open University, 417–61.

Dyer, R. (1977) 'Entertainment and utopia, *Movie* 24 (Spring): 2–13.

Ellis, J. (1982) *Visible Fictions: Cinema, Television, Video*, London: Routledge and Kegan Paul.

Ford, L. (1994) 'Sunshine and shadow: lighting and colour in the depiction of cities on film', in S. C. Aitken and L. E. Zonn (eds) *Place, Power, Situation, and Spectacle: A Geography of Film*, Lanham, Md: Rowman and Littlefield, 119–36.

Friedman, L. D. (ed.) (1991) *Unspeakable Images: Ethnicity and the American Cinema*, Urbana, Ill.: University of Illinois.

Gold, J. R. and Ward, S. V. (1994) '"We're going to do it right this time": cinematic representations of urban planning and the British New Towns 1939–1951', in S. C. Aitken and L.E. Zonn (eds) *Place, Power, Situation, and Spectacle: A Geography of Film*, Lanham, Md: Rowman and Littlefield, 229–58.

Gray, A. (1981) *Lanark*, Edinburgh: Polygon.

Guilbaut, S. (1983) *How New York Stole the Idea of Modern Art: Abstract Expressionism, Freedom, and the Cold War*, Chicago: University of Chicago Press.

Harvey, D. (1989) *The Condition of Postmodernity: An Enquiry into the Origins of Cultural Change*. Oxford: Blackwell.

Herbert, Z. (1985) *Report From the Besieged City and Other Poems*, trans. J. and B. Carpenter, New York: Ecco Press.

Hill, J. (1982) '"Scotland doesna mean much tae Glesca": Some notes on *The Gorbals Story*', in C. McArthur (ed.) *Scotch Reels: Scotland in Cinema and Television*, London: British Film Institute, 100–11.

Hillairet, P., le Brat, C. and Rollet, P. (eds) (1985) *Paris vu par le cinema d'avant-garde: 1923–1983*, Paris: Paris Experimental.

Hirsch, F. (1981) *Film Noir: The Dark Side of the Screen*, New York: Da Cupo Press.

Jameson, F. (1994) 'Remapping Taipei', in N. Browne, P. G. Pickowicz, V. Sobchack and E. Yan (eds) *New Chinese Cinemas: Forms, Identities, Politics*, Cambridge: Cambridge University Press, 117–50.

Kern, S. (1983) *The Culture of Time and Space, 1880–1918*, Cambridge, Mass.: Harvard University Press.

Kracauer, S. (1947) *From Caligari to Hitler: A Psychological History of the German Film*, Princeton, NJ: Princeton University Press.

Leafe, D. and Illott, T. (eds) (1994) *BFI Film and Television Handbook 1994*, London: British Film Institute.

Marx, L. (1980) 'The puzzle of antiurbanism in classic American literature', in L. Rodwin and R.H. Hollister (eds) *Cities of the Mind: Images and Themes of the City in the Social Sciences*, New York and London: Plenum.

McArthur C. (1972) *Underworld USA*, London: BFI/Secker and Warburg.

McArthur, C. (1982) 'Scotland and cinema: the iniquity of the fathers', in C. McArthur (ed.) *Scotch Reels: Scotland in Cinema and Television*, London: British Film Institute, 40–69.

McCann, G. (1990) *Woody Allen: New Yorker*, Cambridge: Polity.

Michelson, A. (ed.) (1985) *Kino Eye: the Writings of Dziga Vertov*, London: Pluto Press.

Natter, W. (1994) 'The city as cinematic space: modernism and place in *Berlin, Symphony of a City*', in S. C. Aitken and L. E. Zonn (eds) *Place, Power, Situation, and Spectacle: A Geography of Film*, Lanham, Md: Rowman and Littlefield, 202–27.

Niney, F. (ed.) (1994) *Visions Urbaines: Villes d'Europe à L'Ecran*, Paris: Centre Georges Pompidou.

Peirce, C.S. (1977) *Semiotic and Significs*, Bloomington: Indiana University Press.

Petley, J. (1978) 'Realism and the problem of documentary', in J. Petley (ed.) *BFI Distribution Library Catalogue*, London, British Film Institute, 3–27.

Petley, J. (1988) 'The architect as übermensch', in P. Hayward (ed.) (1988) *Picture This: Media Representations of Visual Art and Artists*, London: John Libbey, 115–25.

Roffman, P. and Simpson, B. (1994) 'The small town in American cinema', in G. Crowdus (ed.) *A Political Companion to American Film*, New York: Lakeview Press, 395–402.

Rohdie, S. (1969) 'Totems and movies', unpublished seminar paper, London: British Film Institute Education Department.

Rosow, E. (1978) *Born to Lose: the Gangster Film in America*, New York: Oxford University Press.

Sennett, R. (ed.) (1969) *Classical Essays in the Culture of Cities*, Englewood Cliffs, NJ: Prentice-Hall.

Shadoian, J. (1977) *Dreams and Dead Ends: The American Gangster/Crime Film*, Cambridge, Mass: MIT Press.

Short, J.R. (1991) *Imagined Country: Society, Culture and Environment*, London: Routledge.

Sklar, R. (1992) *City Boys: Cagney, Bogart, Garfield*, Princeton NJ: Princeton University Press.

Spignesi, S.J. (1992) *The Woody Allen Companion*, Kansas City: Andrews and McMeel.

Spring, I. (1990) *Phantom Village: The Myth of the New Glasgow*, Edinburgh: Polygon.

Sutcliffe, A. (1984) 'The metropolis in the cinema', in A. Sutcliffe (ed.) *Metropolis, 1890–1940*, London: Mansell, 147–71.

Thompson, K. (1985) *Exporting Entertainment: America in the World Film Market 1907–1934*, London: British Film Institute.

Vernet, M. (1993) 'Film noir on the edge of doom', in J. Copjec (ed.) *Shades of Noir*, London: Verso, 1–31.

Walker, J.A. (1993) *Art and Artists on Screen*, Manchester: Manchester University Press.

Warshow, R. (1962) *The Immediate Experience: Movies, Comics, Theatre and Other Aspects of Popular Culture*, New York: Doubleday.

Wilson, R. (1994) 'Melodramas of Korean national identity', in W. Dissanayake (ed.) *Colonialism and Nationalism in Asian Cinema*, Bloomington: Indiana University Press, 99–104.

Wollen, P. (1992) 'Delirious projections', *Sight and Sound*, August, 24–7.

2

CITY VIEWS
the voyage of film images
•

Giuliana Bruno

"Every story is a travel story – a spatial practice."[1] Film is the ultimate travel story. Film narratives generated by a place, and often shot on location, transport us to a site. Sometimes, the site itself would move. This is the case of Naples, a city that has physically and virtually migrated. Exploring Naples in film, one discovers its bond to New York – a migratory tie, and a filmic *transito*.[2] Situated on the same latitude, Naples and New York are "parallel" places, in many ways adjoined. Several socio-cultural dislocations link these two cities of motion, which are historically and geographically bonded by the cinema.

THE SPECTACLE OF CINE-CITIES

Not all cities are cinematic. Naples and New York are intrinsically filmic. Photogenic by way of nature and architecture, they attract, and respond well to, the moving image. They have housed the cinema from its very first steps, and have a shared history as cine-cities.

The "filmic connection" between Naples and New York is an offspring of their being metropolitan thresholds. These restless visual cities are about motion (and) pictures. As harbor towns, they absorb the perpetual movement of the sea, bear the mark of migration, convey the energy of people's transits, and carry the motion of trade. No wonder the cinema insistently depicts them. They are architecture in motion, as restless as films. An urban "affair", produced by the age of the metropolis, film imparts the metropolitan *transito*, and its ceaseless speed.

Naples and New York lend themselves to moving portraits of dark beauty. Inclined to the *noir*, they portray the metropolis of the near future. Like the filmic metropolis of *Blade Runner*, they are dystopian cities. Always somewhat decayed and embedded in debris, these striking cities are never too far from an exquisite state of ruin. Cities in ruins, they exhibit the social contradictions, and show the high and low, side by side, in the architectural texture, making a spectacle of the everyday. Cities of display, they suit the cinema – the spectacle of motion pictures.

The particular spectacle produced by Naples and New York is street theater. Bonded by the continual motion of their streets, these cities inevitably lead one to the streets to watch the spectacle of their moving crowds. Their filmic image often reproduces this spectatorial walk through town – an urban promenade. New York streets and Neapolitan *vicoli* (alley-ways) have been attracting film-makers for what is now almost a century. The image of these cities, rooted in their street life, is unavoidably linked to the "mean streets." "Escape from New York" and from Naples, is often the message. Yet cinema, and visitors, continue to return to the scene of the crime.

Naples and New York are both tourist attractions, and a tourist's nightmare. Their very history is intertwined with tourism, colonization and voyage, and their relative apparatuses of representation. In many ways, their filmic image partakes in a form of tourism: cinema's depiction is both an extension and an effect of the tourist's gaze. Repeatedly traversed and re-created by the camera, Naples and New York have produced a real tourism of images. Shot over and over again, these cities have become themselves an image, imagery, a picture postcard.

VOYAGE TO NAPLES, A GRAND TOUR

This, of course, has always been true of Naples, the ancient capital, especially renowned for Baroque art and musical life, and for its mix of high and popular culture. A major stop on the Grand Tour of Europe, the elitist touristic practice that preceded mass tourism, Naples was endlessly toured, represented, and pictured. Cinema has continued this type of journey, producing its own grand tour. The ultimate incarnation of the classic voyage to Naples is *Voyage to Italy* (1953–4), a film by Roberto Rossellini (Plate 2.1). In this filmic voyage, as in the Grand Tour, North meets South, and the voyagers' search – a descent – becomes an exploration of the self. Once in the southern environment, the senses erupt, cool equilibriums break, and events take place. A similar experience is described by Jean-Paul Sartre in his letters. Narrating his voyage "down" to Naples, Sartre, as Rossellini, speaks of the city's physicality, finding that one is inevitably affected by the materiality of the place, "the belly of Naples" – a body-city.[3]

Before, and in conjunction with the cinema, images of Naples marked the history of visual arts. Naples was one of the subjects most frequently portrayed in the geography of *vedutismo*, the practice of view painting that was recognized as an independent genre in the late seventeenth century. The image of the city in painting, as in film, is forever linked to the *veduta*, and its representational codes. One wonders whether the Bay of Naples was made to be "pictured."[4]

Naples has been looked at so much that it can be worn out by over-representation. As the writer Thomas Bernhardt once said: "To see the Vesuvius, it is, for me, a catastrophe: millions, billions of people have already seen it."[5] Over-representation and the touristic picture of Naples become a real issue when we consider its filmic portrayal, especially in view of what Naples has been made to

2.1   Roberto Rossellini's *Voyage to Italy*
Courtesy of BFI Stills, Posters and Designs.

symbolize in the cinema. When depicted as folkloric and picturesque, the city becomes both a frozen and a serial image.

The image of Naples in film succeeds a long chain of representations, of which travel literature and the panoramic gaze are an essential part. To understand the connection of Naples to the cinema, one must contemplate the visual history of the city, and consider tourism one of its main motor-forces. The way film emerged in Naples makes this particularly evident. From the very beginning, film was a form of imaging, and a way of touring, the city. Early Neapolitan cinema participated intensely in the construction of "city views." There was even a genre devoted to viewing the city: it was called *dal vero*, that is, shot from real life on location. The first film-makers, including Neapolitan ones, widely practiced *dal vero*, making short films entirely composed of street views or vistas of the city and its scenic surroundings.

Dal vero was an international phenomenon, and Naples became a well-traveled subject of the genre. An extension of nineteenth-century dioramas and panorama painting, *dal vero* speaks of cinema's penchant for "sightseeing." Like the urban spectacle of *flânerie*, the mobile gaze of the cinema transformed the city into cityscape, recreating the motion of a journey for the spectator.

Let us now see how this phenomenon of image-making was implanted in Naples, by touring the origins of cinema in the city, and exploring the migratory connection to New York, established in the early decades of film history.[6]

A PANORAMA OF NAPLES' EARLY FILM DAYS

Naples is the only true Italian metropolis.

(Elsa Morante)

[In Naples] building and action interpenetrate in the courtyards, arcades, and stairways ... to become a theater of the new. ... This is how architecture, the most binding part of the communal rhythm comes into being here, ... [in] the baroque opening of a heightened public sphere. ...

What distinguishes Naples from other large cities is [that] ... each private attitude or act is permeated by streams of communal life.

(Walter Benjamin)

Home of the first Italian railway line, Naples played a prominent role in the rise of another industry of movement: the motion picture industry. Cinema entered the Neapolitan world of spectacle in 1896, and quickly "set down roots." In 1906, the novelist Matilde Serao published a newspaper article entitled "Cinemato-grafeide." The author of *The Belly of Naples* invented this word to name a new kind of virus, a turn-of-the-century epidemic – the cinema. It was attacking the body of her city:

The last and ultimate expression of Neapolitan epidemics and manias, the *dernier cri* of success today is the cinema. Do you see a banner in front of a store? Do not even bother reading the sign. I can swear that it says "Cinema Serena. Coming soon." ... Cinema reigns supreme, and prevails over every-thing![7]

The city's arcade, Galleria Umberto I, served as the main site of cinematographic activities. The implantation of cinema in the arcade was a product of Naples' metropolitan fabric. The arcade was the allegorical emblem of modernity, and, like the cinema, expressed a new urban culture.

This center of commercial, artistic, and social transactions had transposed the spectacle of street life into modern terms. A place of *flânerie*, it had renewed the function of the *piazza* (forum), the traditional Italian site of *passeggiata* (prom-enade), social events, and transitory activities. It was precisely this new urban geography that functioned as catalyst for filmic ventures, together with the area surrounding the railway station. All were sites of transit, indicators of a new notion of space and mobility, signs of an industrial era that generated the "motion picture."

As the home of over twenty film magazines and of a lively film production, Naples was in the vangard of the silent film art's development. This "plebeian metropolis" – as Pier Paolo Pasolini called it[8] – established its own style of film-making. Predating the stylistic features of Neo-realism, it pioneered realistic representation opposed to the aesthetic of "super-spectacles" then dominant in Italy. Neapolitan cinema, rooted in local cultural tradition, was of the street. Often shot on location, it captured the daily popular culture and the motion of street-life.

The three most prominent film companies were Dora Film, Lombardo Film and Partenope Film. All were family enterprises. Partenope Film was run by Roberto Troncone, an ex-lawyer, and his brothers Vincenzo and Guglielmo, an actor. Lombardo Film was a husband–wife collaboration between Gustavo Lombardo and Leda Gys, the star of all their films. Lombardo, an adventurous member of the middle-class who had been involved in Socialist politics, later founded Titanus, one of Italy's most important production houses.

Dora Film was run by a woman, Elvira Notari (1875–1946), and was the most popular of the Neapolitan production "houses." The name and work of Italy's first and most prolific woman film-maker, excised from historical memory, are today unknown. As head of her own production company (named Dora Film after her daughter), Notari made 60 feature films and over 100 documentaries and shorts between 1906 and 1930. She wrote, directed and participated in all aspects of pre- and postproduction and also trained the actors. Her son Edoardo, acting since childhood, grew up on his mother's screen, and her husband, Nicola, was the cameraman. Notari's extensive production, suppressed by Fascist censorship, ended with the advent of sound.

Elvira Notari's city films were particularly sensitive to women's conditions. Shot on location in the *vicoli* of Naples, her (public) women's melodramas derived from the body of urban popular culture. These films, inscriptions of urban transit and panoramic vision, used documentary sequences of street culture and city views. Their narratives reproduced metropolitan topography, the motion of the *piazza* and the geography of the cityscape. Notari's melodrama of the street represented the "belly" of a metropolis, her *meter-polis* – her mother-city, as the word's Greek root suggests.

Notari's films were shown in Naples primarily at a theater in the arcade and endorsed by a passion for the urban travelogue. The Neapolitan film magazine, *L'arte muta* [The Silent Art], remarked in 1916 that film audiences expressed a demand for sightseeing, and expected cinema to satisfy their panoramic desire:

> A moving drama where passion blossoms, red as the color of blood, is produced by *Dora Film*, the young Neapolitan production house, whose serious artistic intentions are well-known. The film was screened at Cinema Vittoria, the movie house of the Galleria Umberto I. The expectations of the public attending this theater have been completely fulfilled: the suggestive drama develops against the enchanting panorama of the city of Naples.[9]

It was, then, the urban matrix of a (plebeian) metropolis that encouraged the circulation and reception of Neapolitan cinema. Naples had traditionally offered the spectacle of movement. Its intermingling of architectural styles created "a baroque opening of a heightened public sphere;" its fusion of dwelling and motion within "a theater of the new," provided a fertile ground for the development of cinema. As film was implanted in the cityscape, the cityscape was implanted within film. Thus transformed into a moving image, the city began to travel.

NAVIGATING FROM NAPLES TO NEW YORK

Elvira Notari's city films were circulated throughout Italy and beyond its borders. They were involved in an interesting process of cultural "migration": the plebeian immigration from Southern Italy at the beginning of the century included "moving pictures." A good portion of Notari's production of the 1920s reached the United States. Like other immigrants, the "immigrant" films by Elvira Notari also traveled by sea, often accompanied by the singers who were to perform the live soundtrack of her musical cinema.

Dora Film found an outpost in New York, a city with a strong Italian immigrant presence, many of whom came from the Neapolitan region.[10] Dora Film of Naples fulfilled the "American dream," becoming Dora Film of America and opening an office in New York City on 7th Avenue. The office functioned as the threshold of distribution. Notari's films circulated most widely among the Italian American community in New York, and reached other North American cities such as Pittsburg and Baltimore, as well as South American countries such as Brazil and Argentina.

During the 1920s Notari's films were screened publicly and regularly in movie theaters in New York's Little Italy and in Brooklyn. Her films were widely advertised in the Italian American daily newspaper, and pompously announced as "colossal Italian cinematography," "directly obtained from Naples and exclusively distributed."[11] The ads were usually quite prominently displayed on the page, and larger than those for performances of *Aida* and *La Traviata* at the opera. Films, operas, and cultural events were mixed together with ads for leisure-time pursuits, vacationing and travel.

Notari's films were generally shown in New York soon after the Neapolitan release, and often before the Roman release. As New York was home to many Neapolitan immigrants, the speed of film "transferral" added a dimension of simultaneous communication to the historical and imaginary connection between the two cities.

When economic and censorial pressures increased in Italy, due to Fascist opposition to the Neapolitan "street cinema," revenues from the immigrant market facilitated the reproduction of these films. The immigrant connection ultimately guaranteed Notari's economic survival through the difficult 1920s, as well as the bypassing of censorship in the first decade of the Fascist regime.

The films arrived in the States with publicity material in Italian and English, including a printed program with stills, illustrations and a summary of the story.

The immigrant films were often described in the poor English of a foreigner. The style of the illustrations recalls the jackets of popular novels, such as those by Carolina Invernizio, a female novelist whose work Notari adapted on to film.[12]

Like the readers of serial popular novels, the viewers of Elvira Notari's films were plunged into the realm of fantasy. The films often spoke of hardship in the popular milieu, and presented fantasmatic strategies of countertactics, as characters negotiated positions of survival in the socio-sexual order at the level of private and public life. They spoke of an experiential condition that their popular immigrant audiences probably knew quite well.

Female spectators were largely present among immigrant audiences.[13] For women, film was a type of leisure approved by society and family. Having replaced previous sites of social interaction, the movie theater had become a place of public acceptance. Film, like the novel, gave women the illusion of social participation while offering the immigrants a sense of fulfilment of personal fantasies.[14] As a form of collective spectacle, film constituted a crucial means of social re-definition for the immigrants. It offered a terrain for negotiating the experience of displacement. As film scholar Miriam Hansen puts it:

> The cinema provided for women, as it did for immigrants and recently urbanized working class of all sexes and ages, a space apart and a space in between. It was the site for the imaginative negotiation of the gaps between family, school, and workplace, between traditional standards of . . . behavior and modern dreams.[15]

The success of Elvira Notari's narrative films among immigrant audiences can be explained in terms of a cultural negotiation between the changing domains of private and public. The private lives of these people had been shaken by the separation from their native lands and milieus. Especially those who came from southern Italian societies, where the *piazza* provided a communal private existence, experienced the shock of a fundamental separation between the private and public spheres in America. Cinema could provide them with an imaginary bond, constructed not only in the actual space of the theater, a site of sociability, but also, and especially, in the mental geography of reception.

Delving into an "ethics of passions," Notari's specialty was the public domain of private passions. At the level of imaging, experience, emotions, sentiments, sexuality – all elements of the private sphere – were heightened. Yet the emotional sphere was "publicly" displayed in the city's space: as part of the theatricality of Neapolitan popular culture, the private dimension would constantly become a social event as the personal was staged in, and for, a public audience. These films thus triggered intense personal/social identification in the social milieu of the movie theater, offering a fictional reconciliation of the split between private and public.

Unlike the immigrants themselves, Notari's films did not travel on a one-way ticket. They completed a cultural and productive round-trip journey: at one end, the American success helped Dora Film to produce films in Italy during difficult times; at the other end, achieving the effect of a cultural travelogue, her films enabled the immigrants to return home. Elvira Notari's films in New York acted as a vehicle to the transformation of memory. In an age of mechanical reproduction, the status of memory changes and assumes a spatial texture. Both personal and collective memories lose the taste of the Proustian *madeleine*, and acquire the sight/site of photographs and films. For the immigrants, Notari's films substituted motion pictures for memory.

THE IMAGINARY JOURNEY FROM NEW YORK TO NAPLES: IMMIGRANT CINEMA

The filmic channel of communication Naples–New York led to an interesting development. Notari's viewers transformed themselves into producers. In the mid-1920s, groups of Italian American immigrants commissioned Dora Film to travel around Italy and make specifically customized short documentary films for them. These shorts were to portray the immigrants' place of origin, the *piazza*, homes, and the relatives they had left behind. According to film historian Vittorio Martinelli, Dora Film produced about 700 of these peculiar travelogues and one feature between 1925 and 1930.[16]

Trionfo cristiano (Christian Triumph, 1930) was the feature commissioned by immigrants from Altavilla Irpina. The sponsors asked for a filmic account of the life of the local patron saint (Pellegrino), who, like Saint Anthony, triumphed over (sexual) temptation. Apparently, a great many naked women performed as temptresses. Unfortunately, both the feature and the shorts appear to be lost or destroyed. If recovered, the immigrants' films would constitute a precious historical archive of a *transito*, of an Italy observed, (re)constructed, transferred abroad – both retained and lost. A cinematic suturing of gaps and differences, these films would be documents, agents, and sources of a *History*. Both a collective History and a personal story, they sustained and supported southern popular cultures displaced from Italy. The immigrant films constituted not only a representation of but also a representation by these subjugated cultures.

Why would Italian immigrants financially sponsor the shorts? The very nature of Elvira Notari's feature films – a fiction that made the "body" of the urban text present as it documented sites and local physiognomies, both recognized and missed by the immigrant viewers – most probably triggered the spectatorial demand for panoramic "home" (made) films. This use of cinema foreshadows the role that moving images have played in the development of video technologies. The shorts functioned as memory snapshots in motion, to have, keep and replay ad infinitum. Before Super-8 and video, they linked cinema's public sphere with the private sphere. Interestingly, and in a way that diverges from the contemporary privatization enacted by the video apparatus, these films were not conceived as private events. They functioned as a personal archive of memories for a community.

The immigrant films were based on a fundamental paradox, in so far as they were travelogues *in absentia*, touristic traces of an impossible voyage. A film itself became a journey, a vehicle of a denied experience – a voyage, so to speak, back to the future, or in the future past. Not a mere substitute for, or a simple duplication of, an event, home movies for people who could no longer go home, whose notion of home was indeed made problematic by emigration, the immigrant films served a socio-historical function by the very fact of their being constituted as private films. The cinematic here and now abridged the immigrants' spatio-temporal distances: it enabled the immigrants to be there, where they no longer were, while remaining here, where they were not yet implanted. The imagery functioned as a shared collective memory while they were in the process of acquiring a new projected identity.

Documents of a cultural history and vehicle for its transcoding into another culture, Notari's films facilitated the process of migration of cultural codes. Participating in the evolution of the immigrants' intersubjectivity, they stimulated a cultural remapping.

While her films continued to reach Italian American audiences, Elvira Notari herself never set foot in New York. As her son stated, she, "despite all invitations, preferred to stay in Naples and send her fans beautiful animated postcards."[17] America left and returned to her as an image: it was a voyage of the imagination. A sequence from a hand-painted film shows a return of the "American dream" in Notari's iconographic universe: a woman is playing a huge drum that bears the inscription "Jazz Jazz Hop Hop Whopee!"

CROSS-CUTTING BETWEEN NAPLES AND NEW YORK: A VOYAGE OF RETURN

Notari's venture had several effects of return. While her production ceased, the immigrant connection continued into the era of sound. Films about Naples and New York began to be produced in New York for the immigrant market. An example of this immigrant cinema is *Santa Lucia Luntana* (1931), also called *Memories of Naples* in English and in Italian *Ricordi di Napoli*.[18] The multiple titles speak of the hybrid status of this film, which has recently been recovered. The stories of Naples and New York, and their languages, intersect in this narrative of immigration.

The film begins in Naples. At the sound of Neapolitan folk music, the spectators are taken for a walk through town. The promenade includes street markets, the Galleria Umberto I and the *piazza* at the entrance of the arcade. Panoramic views of the city and its surroundings abound. City views in the style of *vedute* glorify Naples' renowned vistas. Editing and camera movement reproduce the motion of the city. The mobile gaze of the cinema establishes our presence in the flow of the city. We are comfortably touring the streets and looking at the cityscape. Then, suddenly, there is a jump cut.

In a fraction of time, the jump cut takes us away from Naples, and transports us to New York. In one shot we are in Naples, in the next shot we are in New York. No journey is shown. There is no space or time in between. The motion picture idiom provides the missing link. The experience of emigration is consumed in film editing. It is the void between two shots – an absence, a lack, an off-screen space.

The jump cut erases the pain of the separation, the hardship of the long journey by sea, and the difficulty of the journey into another culture. Suddenly we are in New York, and it does not look so different. A similar film-making style provides the continuity and contiguity that do not exist in reality. We are simply extending our metropolitan journey. As if continuing our promenade through Naples, we end up in New York.

With the moving image leading the voyage, we take another city tour. It begins with the New York skyline, as seen from the water, and proceeds with other metropolitan views. The motion of the city is the protagonist of the excursion. Sites of transit are emphasized as we survey the harbor and visit the railway station. All the way from vistas of Naples to New York's skyline, the cinematic gaze is a tourist's perspective.

From the aerial views of New York, we go down to street level, descending into the city, and enter a neighborhood. A street sign informs us of our whereabouts: East 115th Street. We are in the Italian neighborhood of East Harlem. Visuals and sound provide continuity with Naples. The street looks like an open market and functions like a *piazza*. The documentary style pictures the street market: as we hear the live sounds of the vendors, the camera lingers on the display of food. Everything speaks of a Naples displaced.

In a modest apartment of this area, we are introduced to the protagonists of the story. They are a family of Italian immigrants, composed of an old and disillusioned father, two daughters and one son. Although the mother is dead, she has not left the scene. Her picture, framed on the wall, speaks of her enduring presence. She is the resident Madonna.

The story of *Memories of Naples* is remarkably similar to some of Notari's melodramas. The family dynamics are exactly the same. The film features a generational struggle between an old single parent and a daughter with transgressive socio-sexual behavior. The honest parent, a person of modest means, is also distressed by the actions of an unruly son, who even steals from the family. One of the children provides the consolation of proper conduct. As in Notari's films, the generational struggle speaks of the conflict between an older societal behavior and the demands of the new industrialized world. At issue is the place of woman in modernity, and her public role outside the confines of the home.

In *Memories of Naples*, this conflict is complicated by emigration. America, the New World, is the ultimate representation of modernity. Inhabiting this new "territory" has shown the daughters the way out of a restrictive female role. In the film's view, entering this dimension is quite perilous: by going to work, a good daughter encounters sexual harassment. *Memories of Naples* presents a much more traditional view than Notari's cinema, and is ultimately opposed to modern habits. In its opposition, it collapses different sets of concepts. By treating America, the new land (of immigration), as the New World, it also makes Italy, the old country, into a representative of the old world order. This creates a false idea of Italy, and blurs other distinctions. The generational conflict turns into a clash of cultures, countries and languages. In the film, coming to America has ultimately corrupted the children, who are now, in all senses, estranged from the "universe" of their parents. The father utters bitter words against the new country, that has deprived him of his "youth, dignity," and even "the desire to eat." "America" – he concludes – "is the land of gold and happiness, but for whom?"

Memories of Naples is by no means a singular case, and its typology was not a unique phenomenon. Interestingly, the world view of this Italian-American film is found operating in other types of immigrant cinema. The forgotten world of Yiddish cinema, uncovered by J. Hoberman, presents several cases of family melo-drama, with plot developments that resemble *Memories of Naples*.[19] These films made for the Jewish community also expressed conservative ideas on the family and the New World, especially in the 1930s. A shared master narrative seems to link the Yiddish film to the cinema produced for the Italian immigrants. Furthermore, these two immigrant cinemas shared the same production facilities.

Sound had barely made its appearance in film when *Memories of Naples* was released in 1931. This film is quite remarkable for its use of language and of sound in general. Specific street cries were recorded as emblems of the particular sounds of New York streets. The conflict between the Old and the New Worlds is inter-estingly conveyed by a triadic linguistic structure. In the semiotic universe of emigra-tion, language is a barrier and a hybrid. The film makes simultaneous use of Neapolitan dialect, Italian, and English, with varying degrees of mastery of the latter as a second language. The intertitles are in Italian with simultaneous English translation, while the spoken language is, for the most part, Neapolitan dialect and the English of poor immigrants. Most characters would go back and forth between the two, but there is one person who speaks no English at all. Untouched by the new language, the young man is viewed as uncorrupted. This incontaminate character will turn out to be the family's savior. He will marry the good-natured daughter, and take her and her old father back to Naples.

Marriage provides the resolution of the story. Even the liberated "bad" daughter will end up marrying. No longer a bothersome worry, after the marriage she is literally dropped out of the film at an unknown New York address. As the rest of the family leaves for Naples, the dishonest son is left behind in the East Harlem apartment. The picture of the mother/Madonna also remains, eyeing her son.

Five years elapse. An editing tie covers the distance, and reaches the characters who have returned to Naples. The old father, the couple, and their child are settled in a splendid Neapolitan villa, bought by the repented son, who eventually returned from New York to fulfil the family happiness. The film has now gone full circle.

At the end of *Memories of Naples*, having returned to Naples, we have returned to the beginning of the story – a teasing Neapolitan panorama. An immigrant film has satisfied the immigrants' desire and actualized an impossible voyage of return. This film pretends that emigration is like any voyage: it too has a return destination. In every voyage, the final destination must coincide with the point of departure – home. That which is unattainable in emigration is realized in film. By way of cinema, emigration is transformed into a voyage (home). And, by way of cinema, the journey is completed. It is modernity's travel: the emigration of film images.

ACKNOWLEDGEMENTS

A version of this essay was published in *Napoletana: Images of a City* (ed. Adriano Aprà, Milan: Fabbri Editori, 1993), the catalogue of the film retrospective held at the Museum of Modern Art, New York, 12 November 1993 to 27 January 1994.

NOTES

1 Michel de Certeau, *The Practice of Everyday Life*, trans. Steven Randall, Berkeley: University of California Press, 1984, p. 115.

2 *Transito* (untranslatable in English in one word) is a wide-ranging notion of circulation, which includes passages, traversals, transitions and transitory states, and incorporates a linguistic reference to transit. See Mario Perniola, *Transiti: Come si va dallo stesso allo stesso*, Bologna: Cappelli, 1985.

3 Jean-Paul Sartre, *Witness to My Life: The Letters of Jean-Paul Sartre to Simone de Beauvoir 1926–1939*, ed. Simone de Beauvior, trans. Lee Fahnestock and Norman MacAfee, New York: Charles Scribner's Sons, 1992.

   The commonly used expression "the belly of Naples" ("*il ventre di Napoli*") derives from a book by the female novelist and journalist Matilde Serao. See Matilde Serao, *Il ventre di Napoli*, Naples, 1884.

4 On Naples and the Grand Tour see Cesare de Seta, "L'Italia nello specchio del Grand Tour", in *Storia d'Italia*, vol. 5, Turin: Einaudi, 1982. On *vedutismo* and Naples, see *All'ombra del Vesuvio: Napoli nella veduta europea dal Quattrocento all'Ottocento*, Naples: Electa, 1990 (exhibition catalogue).

5 Cited in Fabrizia Ramondino and Andreas Friedrich Müller, eds, *Dadapolis: Caleidoscopio napoletano*, Turin: Einaudi, 1989, p. 181.

6 For an extensive treatment of this subject see Giuliana Bruno, *Streetwalking on a Ruined Map: Cultural Theory and the City Films of Elvira Notari*, Princeton, NJ: Princeton University Press, 1993.

7 Gibus, "Cinematografeide", *Il Giorno*, March 30, 1906. Reprinted in Aldo Bernardini, *Cinema muto italiano*, vol. 2, Bari: Laterza, 1981, pp. 20–1. Gibus was Matilde Serao's pseudonym.

8   Pier Paolo Pasolini, *Lettere luterane*, Turin: Einaudi, 1976. See in particular p. 17.

9   *L'arte muta*, Naples, vol. 1, no. 1, June 15, 1916.

10   See Thomas Kessner, *The Golden Door: Italian and Jewish Immigrant Mobility in New York City, 1880–1915*, New York: Oxford University Press, 1977; and Louise Odencrantz, *Italian Women in Industry*, New York: Russell Sage Foundation, 1919.

11   See, for example, *Il Progresso Italo-Americano*, July 9, 1921 and Sept. 1, 1924.

12   Invernizio's novels were distributed in the States by a New York-based Italian American Press, located at 304 East 14th Street in Manhattan.

13   See Roy Rosenzweig, *Eight Hours for What We Will: Workers and Leisure in an Industrial City, 1870–1920*, London and NY: Cambridge University Press, 1983, in particular, pp. 190–215; Elizabeth Ewen, "City Lights: Immigrant Women and the Rise of the Movies," *Signs: Journal of Women in Culture and Society* vol. 5, no. 3 (supplement) 1980, pp. 45–65; idem, *Immigrant Women in the Land of Dollars*, New York: Monthly Review Press, 1985; Kathy Peiss, *Cheap Amusements: Working Women and Leisure in Turn-of-the-Century New York*, Philadelphia: Temple

University Press, 1986; and Charles Musser, *Before the Nichelodeon*, Berkeley: University of California Press, 1991.

14   This argument is developed by Judith Mayne. See her *Private Novels, Public Films*, Athens and London: The University of Georgia Press, 1988, in particular the chapter "The Two Spheres of Early Cinema."

15   Miriam Hansen, *Babel and Babylon: Spectatorship in American Silent Film*, Cambridge, Mass: Harvard University Press, 1991, p. 118.

16   Vittorio Martinelli, "*Sotto il sole di Napoli*," in *Cinema & Film*, vol. 1, eds Gian Piero Brunetta and Davide Turconi, Rome: Armando Curcio Editore, 1987, p. 368; and the interview published in *La Repubblica*, January 29, 1981, p. 16.

17   Interview with Edoardo Notari, cited in Stefano Masi and Mario Franco, *Il mare, la luna, i coltelli*, Naples: Pironti, 1988, p. 160.

18   A Cinema Production, Inc. (president S. Luisi); director: Harold Godsoe; cine-matographer: Frank Zucker; studio: Forth Lee, New Jersey; producer: Angelo De Vito; music: Giuseppe de Luca.

19   J. Hoberman, *Bridge of Light: Yiddish Film Between Two Worlds*, New York: Pantheon Books, 1991.

3

OF PLANS AND PLANNERS

documentary film and the challenge of the urban future, 1935-52

●

John R. Gold and Stephen V. Ward

The philosophy is education, you see. But if you don't have a little entertainment and action, you don't get the opportunity to tell your story.
MARLIN PERKINS, *quoted in Wilson, 1992: 134*

INTRODUCTION

In the early years of the Second World War the Ministry of Information commissioned the London-based film-production company, Strand Films, to make a short documentary about the need for town planning. Entitled *New Towns for Old* (1942) and having a script written by Dylan Thomas, the film's action is framed around the dialogue between two bowler-hatted actors as they walk around the Northern steel-town of 'Smokedale' (in reality Sheffield). One, speaking with a 'no-nonsense' Yorkshire accent, dispenses the knowledge of the insider. He imparts practical insight into 'how folk live here', although he also stresses that 'they live like this in most . . . other big towns'. Throughout the six-minute film, he extols the virtues of planned action to rebuild the industrial city, in particular arguing that towns should be planned so that residential and working areas are separated from one another.

His companion, who has a Southern English accent, acts as a foil. As an outsider, he continually expresses polite doubt and asks to be shown evidence for what he is being told. Gazing at the city's landscapes, he is incredulous to hear that clearance should be employed to separate out different land-uses: 'But you can't move a town around like that.' His associate disagrees. Halting outside a building that displays a large and conveniently symbolic sign stating 'Public Cleansing Station', he invites the visitor, and the audience, to 'look down there' at the vista of a seemingly devastated residential area. Asked if the houses had been flattened by bomb damage, he replies:

No, we pulled them down ourselves. These were slums worse than the ones you just came through. You see, we can replan a town if we want to and we have planned it an'all.

The radical thrust of these comments may seem remarkable to modern observers given the conservative influences that surrounded the film's production. The need to operate within the restrictions of wartime censorship and to accept sponsorship from an arm of the state officially responsible for propaganda 'to present the national case to the public at home and abroad' (Aldgate and Richards: 1986: 4) might seem the recipe for tacit conformity to the status quo. Yet the sentiments expressed would have neither surprised contemporary film audiences nor would have been regarded as unduly contentious. There was a national sense of gratitude to the people of Britain's cities for their steadfastness in the face of the Blitz.[1] The wish to promise a better, planned future carried much the same emotional appeal as had the slogan 'homes fit for heroes' in the First World War. It was what the people deserved in view of what they had endured. In addition, a broad consensus had developed around the necessary strategies for tackling the problems of Britain's housing and urban infrastructure after the war. The cessation of house-building, suspension of the slum clearance programme, lack of investment in transport and services, and damage from aerial bombardment created circumstances that required action on a national scale. Planning in general, and town planning in particular, were an essential part of the equation.

Elsewhere, we have supplied a chronological analysis of the ways in which British documentary film-makers contributed to the debate about planning between 1939–51 (Gold and Ward, 1994). From initial concerns with housing and the slum clearance drive, film-makers progressed to championing the broader cause of town-planning, before gradually moving on to giving support to the idea of the Garden City/New Town as a specific prototype for the future. Here, we focus specifically on the way that documentary film-makers depicted, first, the planning process itself and, second, the planner as the mediator of a better urban future. Drawing on a representative selection of films released between 1935 and the demise of the British documentary movement in 1952,[2] we examine the contrasting ways in which town planning and the professional town planner were presented to cinema audiences.

This chapter contains five main sections. The first outlines the nature of the city in film and considers the reasons for the British documentarist's engagement with the city and its problems. The second part traces the original preoccupation with housing *per se* as an extension of reformist debate about health and sanitation, before noting the emergence of films that recognised that housing reform needed to be placed in the context of town planning. In the third section, we outline the search for engaging ways to present the technical activity of planning, identifying a recurrent set of themes that film-makers employed. The fourth section deals with the image of the 'planner' to be found in these films, varingly depicted as a visionary, scientist, member of a team or anonymous technician. We make particular reference here to depictions of Sir Patrick Abercrombie, the leading town planner of the 1940s, who made a prominent appearance in several documentaries of the period. The conclusion reflects on the challenge faced by documentary film-makers in promoting understanding of the important, but cinematically dull, activity of town planning to the lay public, noting how their work provides a sensitive guide to wider thinking about the nature of town planning.

THE DOCUMENTARY TRADITION

THE CITY IN FILM

Cinematic film has long had a close and many-sided relationship with the metropolitan city. Leaving aside the economic factors that drew the commercial cinema to major centres of population and the growing appreciation of cinema buildings as important built forms in their own right (e.g. Richards, 1984), the most important aspects of that relationship are reflected in film content.

Many commentators (Solomon, 1970; Wiseman, 1979; Sorlin, 1991) have argued that film itself developed as an 'urban art', frequently articulating its narratives against the backdrop of the metropolitan city. This was partly due to the traditional location of the major film studios in and around such major cities as Los Angeles, New York, London, Paris, Berlin and Rome. Early film-makers were intrigued by the possibilities that their varied and ever-changing neighbourhoods offered as settings for fictional film. Yet while their films contained many powerful images of city scenes – Harold Lloyd hanging from the hands of a clock or Keystone Cops chasing through the suburbs of Los Angeles – the setting did not unduly influence the psyches of the human participants in this phase of the silent cinema. The city, as Larry Ford (1994: 120) notes, was just 'there'.

This changed during the interwar period. Films made in Berlin and Hollywood in the 1920s started to reflect the intellectual hostility to the city that already existed in literature, painting, theatre and the humanities (Lees, 1985; Carey, 1992). Fascinated and repelled by the city in equal measures, film-makers developed new imageries. Cities were now typically enclosed, overcrowded, noisy and tense. Where necessary, the impression of oppressiveness would be reinforced by juxtaposing such scenes with romanticised images of a bucolic countryside (McArthur, this volume). Though a small minority of film-makers championed the city's aesthetic and cultural virtues, the majority set their stories of crime, passion and intrigue in cities polarised by enormous contrasts in living conditions and characterised by a seamy underlife all too ready to rise to the surface (Gold, 1984, 1985: 125).

Such scenes, however, were rarely shot on location. Due partly to the arrival of sound and the technical restrictions that the new medium imposed, film-makers rarely left the confines of the sound-stages or exterior sets constructed within the grounds of the feature studios. The creation of the sinister and hostile city, therefore, was either carefully crafted by set designers or else remained mostly implicit, fleetingly glimpsed through the windows of restaurants or speeding cars. It was less a place than a state of mind (Ford, 1994: 121). Only in the late 1940s, with the rise of neo-realist cinema, were film-makers encouraged to believe that the same atmospheric scenes could be created as effectively on location (Webb, 1987: 6–7).

DOCUMENTARISTS AND THE CITY

Given this background of film practice and ideology, it is not surprising that the documentary movement took some years to define its own position on the city and its problems. In appreciating that history, it is important to stress that documentary and 'fictional' film are not polar opposites. Although documentary films are usually defined as films distinguished by a high factual and sociological content and a subject matter that charted the lives of people in their everyday environments, they share many characteristics with 'fictional' film. Either genre can be constructed around a narrative decided by the film-maker, or have educational intent, or employ locational work in real-world cities. Perhaps the most significant difference lies in the *mise-en-scène*, the director's control over what appears in the film frame. In controlling the *mise-en-scène*, the fictional film director stages the event for the camera, whereas many would argue that the documentarist does not rely on *mise-en-scène* (Bordwell and Thompson, 1993: 145).

Applied to the present context, it may be observed that many of the earliest uses of film to create 'documents' that record the activities of people going about their everyday lives were shot in European and American cities at the turn of the twentieth century (e.g. see Macfarlane, 1987). Over time, new images of the city and urban life developed as the documentary film itself evolved into various new forms (Thompson and Bordwell, 1994: 344–68). Montages of urban scenes in *cinéma-vérité* style, for example, emanated from the Russian and German cinema.[3] Municipalities engaged in film-making activities for promotional purposes.[4] Progressive, or Leftist, film-makers drew attention to poor living conditions and to cities as crucibles of revolutionary protest.[5]

The mainstream documentary movements in the USA and Britain, however, generally neglected the city until the late-1930s. In the USA, the prime reason was the urgency of an alternative focus. Films such as *Our Daily Bread* (1934) by King Vidor and *The Plow that Broke the Plains* (1936) and *The River* (1937) by Pare Lorentz examined instead the problems of rural America. Lorentz, for example, charted the social and economic history of the Great Plains from their settlement through to the years of Depression and drought. Inspired by Soviet experimental cinema, his films pointed to the severe problems caused by the operations of sharecropping, soil exhaustion, unchecked erosion, flooding and subsequent migrations (Alexander, 1981; Snyder, 1993). Only with the showing of *The City* (1939a) at New York's World's Fair was the public presented with documentary images of American cities and their problems (see p. 69).

In Britain, belated portrayal of the city also reflected the pattern of development of the documentary movement and its characteristic concerns. This was heavily influenced by the work of the Scottish film-maker John Grierson. Credited with coining the word 'documentary' in 1926,[6] Grierson had spent his formative years in the USA where he was impressed with the cinema's power to reach mass audiences but deplored 'the way Hollywood cinema missed its opportunity to combine entertainment with education' (Thompson and Bordwell, 1994: 354). Through his efforts as director and administrator, Grierson gave impetus to a movement that produced more than one thousand films between 1929 and 1952. While detailed consideration of that contribution lies beyond the scope of this essay,[7] three points are significant in understanding the particular concerns of that movement.

First, Grierson emphasised 'public service'. Characterised as 'a decidedly non-Marxist young radical with a strong belief in the educational potential of film' (Hogenkamp, quoted in Armes, 1978: 128), Grierson believed that documentary film could serve important didactic functions. The documentary film offered a window on the world. Developments in film technology made it both practically possible and economically feasible to move out from the confines of the studios and to 'photograph the living scene and the living story. . . . [The] original (or native) actor, and the original (or native) scene, are better guides to a screen interpretation of the modern world [than are studio films]' (Grierson, quoted in Williams, 1980: 17).

Second, while documentary films sought to present the truth about the real world, there was no assumption that they would passively mirror it as some have asserted (e.g. Armes, 1978: 134). While Grierson never seriously questioned the idea that the camera never lies, he also argued that truth was an interpretation or perception that would be revealed only when the film-maker had arranged the subject matter into a suitable form (Aitken, 1990: 7). Truth was not produced by simply turning on a camera and pointing it at an appropriate subject, but emerged from the creative notions that guide the various stages of film production from preparation, through shooting, to assembly.

Third, to tell that truth, documentarists followed the broad ideological leads provided by Grierson. These stressed both aesthetics and the sociological purpose of film-making. Aesthetically, documentary films went beyond merely compiling information to construct 'a visual art' that could convey 'a sense of beauty about the ordinary world, the world on your doorstep' (Aitken, 1990: 11). Indeed some argued that the British documentarists' innovations in sound and picture editing represent their major contribution to film history (e.g. Barsam, 1989). Rather more, however, would emphasise their sociological contribution in producing films that highlighted the social and economic conditions of the 1930s and 1940s.

When tackling such matters they were initially restricted by the twin constraints of sponsorship and censorship.[8] To elaborate, while fringe film-makers were free to apply radical political analyses in their films about the slums in the sure knowledge that their films would scarcely be seen outside the circles of the committed, their colleagues in the mainstream documentary movement perforce trod more carefully. Documentary film finance came from commercial and state sponsors. While the evidence suggests that sponsors seldom intervened directly in film direction or content, most would have found a position of loose social-democratic reformism inherently more palatable than overt radical criticism of the social order (Aitken, 1990: 9).

Censorship also imposed limitations. The British Board of Film Censors (BBFC) imposed a highly detailed set of rules throughout this period that firmly discouraged the cinematic treatment of any controversial social or political issues (Pronay, 1982). In the early 1930s, for example, would-be producers of a film version of Walter Greenwood's novel *Love on the Dole* (1933) were warned not to proceed by the BBFC because it would show 'too much of the tragic and sordid side of poverty' (Richards, 1984: 120). This attitude did not change decisively until the war years. Even in 1937, when the documentary movement was actively challenging the boundaries of cinematic acceptability, the President of the BBFC, Lord Tyrrell, could proclaim publicly his great pride that 'there is not a single film showing in London today that deals with any of the burning issues of the day' (cited in Pronay, 1982: 122). The longstanding dominance of such attitudes obviously discouraged any particularly penetrating cinematic treatments of socially contentious issues such as the slum problem. Initially the slum clearance drive initiated by the Labour Government (1929–31), with its implications of positive action to tackle inequality, would have come into the 'controversial' category. It was only when all-party consensus emerged in the mid-1930s that a spate of documentary films appeared on the subjects of slum clearance and rehousing. Even then, the focus was not so much on the slum problem as on the solution of slum clearance and redevelopment.

BEYOND HOUSING

Newsreels had often contained positive images of poor housing conditions being briskly solved by slum clearance since 1930,[9] but fuller-length documentary treatment dates from the release of Arthur Elton and Edgar Anstey's *Housing Problems* (1935). It was filmed on location in East London although it also incorporated silent footage, dating from 1928, drawn from the archive of the London Borough of Bermondsey's Health Department (Lebas, 1992). While its style was directly influenced by that of the American newsreel series *The March of Time* (Fielding, 1978), *Housing Problems* broke new ground by its use of face-to-camera interviews. Through interviews with residents of slum housing and with people newly rehoused in a model block of flats (Lea View House, Hackney), the film maps the progress of people from despair to hope. Yet despite this familiarly optimistic storyline, the film also projects a real note of social criticism. It provides a revealing portrait of the interlinkage between poverty and wretched living conditions and of the broad social responsibility to ensure improvements. The film also displays willingness to edge away from characteristic interwar uncertainty as to 'whether the slum dwelling or the slum dweller produced unhealthy living conditions' (Garside, 1988; see Gruffudd, 1995: 42). Here notions of public health are directly related to environmental reform.

The content and, to some extent, the form of *Housing Problems* were copied elsewhere. The closest resemblance is found in *The Great Crusade: the story of a million homes* (1937), essentially an officially approved version of its predecessor. Made by Pathé for the Ministry of Health, it places a cheerful face on the National Government's scheme to eradicate slums in Britain by 1938. Other films cover particular aspects of the same territory. John Grierson's *The Smoke Menace* (1937), Frank Sainsbury's *Kensal House* (1937) and Paul Rotha's *New Worlds for Old* (1938) all chart the decrepit and verminous state of overcrowded slum housing and chronicle the progress of clearance schemes. Purpose-built working-class flats generally receive favourable comment as a symbol of progress that provides hope for the future. It is interesting, however, to note what is omitted (Marwick, 1974). No film-maker ever doubts that slum-dwellers are anything other than delighted with their new homes although complaints that would become familiar in the post-war years – about noise and the problems of living with children in multi-storey flats – had already begun to be voiced. Equally no-one directly attributes blame for causing the slums in the first place other than saying that they were the result of history and unenlightened practices. Finally, none of these films deals directly with the more general question of town planning.

Perhaps the earliest formal attempt to rectify that last deficiency can be found in *Housing Progress* (1938). This 20-minute silent two-reeler was produced by Matthew Nathan for the Housing Centre – an independent body founded in 1934 to provide housing information, publicity and research. It begins with a conventional review of the nature of the housing problem followed by a commentary on the architectural styles of modern flats in London, such as the Ossulston estate (St Pancras), Princess Alice House (Kensington) and the newly built Kensal House in Ladbroke Grove. Nevertheless, after cataloguing the successes of rehousing using flats, the film indicates that other problems would remain unless attention was paid to the wider context. One caption notes:

> But the building of many-storied blocks of flats, when it is done piecemeal, creates high densities and will result in congestion, traffic problems and the obstruction of light and air. For this reason the Housing Act contains special provisions to encourage the redevelopment of large central areas according to a comprehensive plan. Liverpool Corporation, the London County Council and Norwich Corporation have led the way with 'redevelopment areas'. Will other local authorities follow?

The construction of cottage housing estates in outlying areas also posed problems. Reviewing the work of the London County Council in building estates beyond the county boundary at Becontree, St Helier, Downham and Watling, the caption-writer remarks that: 'such dormitory estates are often lacking in social amenities and as long as work remains centralised the tenant will have to make long and expensive daily journeys, and traffic problems will be increased.' The remedy canvassed by the film-makers was for construction of satellite towns: 'The logical solution is the decentralisation of industry and the creation of planned garden cities or satellite towns comprising an industrial and residential quarter and social amenities.' Favourable mention is then made of the examples set by the Garden Cities at Letchworth and Welwyn and of Manchester Corporation's Garden City-inspired estate at Wythenshawe.

Given that it was a low-budget instructional film, *Housing Progress* received few showings outside the realm of the film societies. Despite this, it may be seen as a useful index of two important developments. First, it presages the movement to a propagandist style of documentary that, as we have shown elsewhere (Gold and Ward, 1994), specifically canvasses the merits of the Garden City as a prototype for the urban future. Second, it clearly argues that effective slum clearance also requires that attention be paid to town planning, even if there is little explanation about the processes involved. There was not long to wait, however, before films addressing this issue started to appear.

REPRESENTATIONS OF PLANNING

The need to find engaging ways to present town planning to cinema audiences took on greater importance with the onset of war and the need to think about post-war reconstruction. Despite the passage of various Town Planning Acts during the twentieth century, town planning remained ill-defined and poorly understood by the general public. Its outcomes, expressed in the two-dimensional language of maps and plans, lacked cinematic impact when compared, say, with the visually compelling examples of three-dimensional forms that could be shown as the end-products of architecture.[10] Film-makers, therefore, had to find strategies to present town planning as a more vivid and readily intelligible area of endeavour. Over time four such strategies were employed. Three of them – planning as the application of science, as social medicine and as revelation – were present even in the housing documentaries and frequently overlap. The fourth – planning as wizardry – arose primarily due to the failure of the other strategies to present planning in a readily interpreted manner.

SCIENCE AND RATIONALITY

The 1930s was a decade in which many championed rational philosophy and the socially-redeeming virtues of science and technology to cure the problems of society. From that perspective, a comprehensive and scientifically based system of town planning was naturally favoured as a way of allocating resources and bringing about social improvement. Both *Housing Problems* and *The Great Crusade* saw film-makers presenting images of the advanced constructional techniques employed in building flats as an example of scientific methods for housing improvement.

Other films extended that analysis to planning the city as a whole. Alberto Cavalcanti's film *The City* (1939b), commissioned by the General Post Office, presents a full-scale analysis of London's problems by the architect–planner, Sir Charles Bressey, followed by an exposition of radical proposals for improvements to aid the functioning of London's transport system. *New Towns for Old* (1942), mentioned above, provides brief analysis of the application of zoning in urban reconstruction, accompanied by a compilation of nineteen different scenes, rapidly intercut, to convey an impression of both the complexity and the frenetic pace of planning activity. The Bournville Village Trust's film *When We Build Again* (1943) brought the interesting innovation of a short acted sequence, involving three serving soldiers returning home on leave, to explain the zoning pattern of Birmingham before drifting into a standard exposition, through models and overlays, of planning ideas. The latter includes how new settlements could be scientifically redesigned using modern ideas on zoning, neighbourhood planning and the design of road networks.

These themes were taken further in *Proud City* (1945), which provides the most consistent cinematic discourse on planning as applied science. Commissioned by the Ministry of Information, *Proud City* seeks to explain the thinking behind the County of London Plan (Forshaw and Abercrombie, 1943), a document that embraced many principles that became a standard part of post-war planning.[11] In doing so, the film presents planning as a scientific, rational and empirically based process and those responsible for preparing the Plan (members of the London County Council's Architect's Department) as the technicians in charge of that process.

The film opens by reiterating the commonly expressed view that the damage caused by war should be seen as a historic opportunity not just to repair London but also to undertake its reconstruction. This is presented as the will of the people. Voice-overs by the 'people of London' indicate their wishes for decent housing, parks, less traffic problems and safer roads. The views of business or interest groups including the Metropolitan Police, the Port of London Authority, the Water Board and private industry are indicated by re-enacted or simulated conversations. Seen against this background, planning is portrayed as a progressive but reactive enterprise, which identifies the needs of Londoners, resolves any conflicts, and devises an appropriate plan.

The first stage in the process is to undertake an exhaustive survey: 'first of all we had to find out everything about the great city we were planning to rebuild. Everything about its history, its geography and the way that they [i.e. Londoners] live.' Rapid cutting from scene to scene supplies a visual summary of the range of survey activities undertaken and creates an impression of programmatic activity. Surveyors are seen using theodolites and field-workers attentively write down the opinions of their respondents. The ensuing analysis proceeds in a busy drawing office. Members of the Architect's Department pore over charts, consult one another over diagrams or sit alone working on maps with their T-squares. The audience are told that 'many thousands of maps and charts were drawn, modified, discarded, revised and done again' in the painstaking preparations that lay behind the County of London Plan.

3.1   Arthur Ling, in the guise of an anonymous technician, explains the ideas of neighbourhood planning found in the *County of London Plan* (1943). *Proud City* (1945).
© Crown Copyright Films. Courtesy of BFI Stills, Posters and Designs.

The next, and longest, part of the film sees the proposals outlined. J. H. Forshaw, the LCC's Chief Architect, and Sir Patrick Abercrombie in tandem (see p. 74) awkwardly describe the plan's main precepts. It is introduced as a rational answer to problems posed by London's rapid growth and resulting sprawl. Above all, they argue that the survey and resulting analysis had successfully uncovered the old village communities of London, showing that 'the old local loyalties are still there'. The plan offers the chance to equip the built fabric of London with a new cellular structure to match its social fabric.

Arthur Ling, then a member of the LCC's Architect's Department, furnishes a more specific explanation of the principles for planned reconstruction (Plate 3.1). After identifying areas where immediate clearance and rebuilding might take place, he begins his exposition with an example of the design for a social area or 'neighbourhood unit' planned for Stepney. Using charts and then a three-dimensional model, he illustrates the design features incorporated into a single neighbourhood unit, then shows how neighbourhood units would be linked into the wider whole (see also Gold, 1995). At city level, Ling outlines plans for remoulding the road and rail patterns for London and for regenerating London's waterfront. Overall, a strong impression is given of the power of rational, scientifically based planning, in which painstaking analysis first identifies the basic socially defined building block for the reconstruction process and, second, provides a consistent set of axioms through which to create the new city.

SOCIAL MEDICINE

The notion of planning as applied science was soon supplemented by other strategies. One of the most significant was the idea of planning as social medicine. Precedent for this was also found in the housing documentaries with their preoccupations with health and sanitation, their powerful images of vermin and of fungus growing on damp-infested walls, and their suggestion that positive action would improve both physical and social health. A documentary originally produced for showing at New York's World's Fair (1939), but widely seen by British film audiences in the early war years, develops this further by propagating an organic approach that sees planning as a means to preserve the health of both the city and its society.

Commissioned by the American Institute of Planners with the aid of a $50,000 grant from the Carnegie Corporation of New York, *The City* (1939a) is based on a scenario derived from Lewis Mumford's seminal text *The Culture of Cities* (1938).[12] Containing commentary by Mumford himself and a powerful musical score by Aaron Copland, the film's opening caption states: 'Year by year our cities grow more complex and less fit for living. The age of rebuilding is here. We must remould our old cities and build new communities better suited to our needs.'

In reality, however, the film builds a stronger case for the use of planning to create new communities or satellite towns on greenfield sites rather than to remould existing cities. The film has a three-part structure. The first paints a glowing portrait of life in a New England village. Residents practise timeless crafts, children wander freely and a town meeting emphasises the democratic basis of a community where all can participate. The second section assembles a montage of scenes of urban and industrial life to argue that increasing scale and the dominance of the machine had robbed the people of their inheritance. Images of pollution, dirt and grime, poor housing, and children playing in conditions of great danger testify to the impoverishment of life for the urban masses. Lengthy scenes of jostling crowds and traffic chaos, reminiscent of King Vidor's silent Hollywood classic *The Crowd* (1928), and a famous scene of people eating mechanically in an automated café, indicate the triumph of the machine over human life. Nevertheless, technology is a two-edged sword that also offers the chance for people to reassert control over their living environments if used properly. Indeed, it could even help to restore the harmony between people and the land. Against a background of images of new technological marvels, Mumford intones the following:

> This new age builds a better kind of city, close to the soil once more. As moulded to our human wants as planes are shaped for speed. New cities take

form, green cities. They are built into the countryside. They are ringed with trees and fields and gardens. New cities are not allowed to grow and over-crowd beyond the size that makes them fit for living in. The new city is organised to make cooperation possible between machines and man and nature.

The remainder of the film shows scenes depicting the intervention of planning to restore the health of urban communities. Using footage taken from five different locations[13] but 'put together to make a synthetic community' (Levin, 1971: 189), the new 'green city' comprises low density housing in green surroundings, with roads designed to allow accessibility but to keep through-traffic moving smoothly. Industry is accommodated in modern buildings, separate from residential areas but close enough for workers to walk or cycle to their place of employment. Families stroll together or attend open-air concerts. Above all, children can again roam and play freely in green surroundings, taking part in both impromptu and organised outdoor games. They bathe in the lake or cycle into the woodlands to picnic with their friends in the open air. While the screen shows scenes of new sanatoria and modern schools, shot in the characteristic heroic style of 1930s' black-and-white cinematography, the narrator argues that: 'City and school and land in active partnership provide the raw materials for life and growth. Here boys and girls achieve a balanced personality ready to build and meet a many-sided world.' The healthy growth of children is essential for the healthy growth of the community. Good planning nurtures the coming generation, affords them a decent start in life and restores the connection between the people and the land. In keeping with the themes of organicism and social health, town planning is depicted as a powerful force for social medicine.

THE PURSUIT OF VISION

Another strand in the representation of planning was added by Alberto Cavalcanti's film, confusingly also entitled *The City* (1939b). Its main content is an illustrated lecture by Sir Charles Bressey, an architect–planner who had recently jointly authored a report on London's transport problems with Sir Edwin Lutyens (Bressey and Lutyens, 1937). While the notion of planning as science is important, another element – the idea of planning as the pursuit of vision – is also put forward.

It is introduced by considering Sir Christopher Wren's plan for rebuilding the city of London after the Great Fire (1666). After noting that fire had cleared away the masses of wooden houses that had stood for centuries, the camera then cuts to a portrait of Wren. When panning back, it is seen to be hung on the wall of Bressey's office. Bressey extols Wren as someone who 'alone was wise enough to see the great opportunity'. He praises 'the far-sighted nature' of Wren's plan for a London of 'broad sunlit avenues, well-placed monuments and public buildings', but states, with a sigh, that it was never built: 'The forces of conservatism were too strong.'

As the film progresses, the audience is shown many images of the unsatisfactory state of contemporary London, particularly concerned with its persistent and worsening traffic congestion. Here a parallel is drawn between Wren and Bressey. The narrator notes that: 'Just as Wren made a plan for seventeenth-century Londoners, so Sir Charles Bressey has worked out a plan for us.' Bressey and Lutyens' proposals had envisaged a new system of inner and outer orbital roads, elevated cross-city roads, tunnels, multi-level roundabouts and clover-leaf intersections. Yet the visionary is again thwarted. Morosely gazing at the traffic out of his office window, Bressey regrets that it would take an hour for him to journey home this evening. Had his plan been implemented, he need not have spent more than fifteen minutes.

Not all films, however, convey frustration at the rejection of visionary schemes. The war dramatically increased the salience of planning so that even the most idealistic projects no longer seemed unattainable (see Ward, 1994: 80–99). Film became a powerful vehicle in the popular articulation of this vision, reflecting and reinforcing planning's enhanced importance. *Town and Country Planning* (1946), made by the Army Bureau of Current Affairs for the War Office, looks forward to dramatic improvements in the urban environment through the agency of comprehensive planning and, particularly, through a National Plan for housing in which central and local government would closely cooperate. Jill Craigie's *The Way We Live* (1946) affords a somewhat deeper analysis of visionary notions for the reconstruction of Plymouth contained in the *Plan for Plymouth* (Paton Watson and Abercrombie, 1943). Using a mixture of acted scenes, stock footage, new documentary sequences and some personal appearances by Abercrombie himself (see p. 74), *The Way We Live* charts the origins and anatomy of the Plan, coupled with an indication of local response. While the film-makers give every impression of sympathy with the 'progressive' case of the plan-makers, they also point to the intensity of local objections.

Perhaps the most striking, and radical, expression of this theme is found in Paul Rotha's *Land of Promise* (1946). Structurally it contains three parts, respectively, homes 'as they were', 'as they are' and 'as they might be'. The straightforward analysis of inner-city housing problems in the first two segments develops into a discourse on the benefits of a planned society in the third. Town planning, it is argued, needs to embrace new technology and the opportunity presented by the post-war situation to act as an important force for change, helping to bring about social planning, 'economic democracy' and the abolition of the profit system (Marris, 1983). One scene, for example, has two characters off-camera debating the future.[14] In a manner reminiscent of *New Towns for Old*, one expresses optimism about the possibilities for change presented by the circumstances of the time, the other is deeply sceptical. The evidence for optimism is conveyed by an animated commentary delivered as voice-over to a rapidly intercut montage of archive footage of constructional activity. Reconstruction is compared to waging war. Just as new methods allowed the rapid construction of wartime airfields, they could do the same for new housing, roads, and towns. A composite vision is supplied of a new community equipped with 'health centres, clinics, libraries, nurseries, parks, cinemas and theatres; the whole bag of tricks located and designed according to a proper plan.'

PLANNING AS WIZARDRY

The pursuit of vision was added to the other strategies as part of the continuing effort to find engaging and illuminating ways to present planning to a lay audience. Problems, however, persisted. Despite the film-makers' best endeavours, planning came across as a worthy but somewhat dull activity, explained by practitioners who recited their lines lifelessly to camera. As time went by, more films turned to using actors and simplified scenarios to communicate their meaning, with progressively less attempt to convey the complexity and technical detail of town planning.

This theme is amply exemplified by a series of promotionally oriented documentary films publicising the first-generation of new planned settlements created under the 1946 New Towns Act.[15] All the familiar problems of portraying what planning could do in the future in advance of what currently existed became more acute in the selling of the New Town. The post-war Labour Government were also anxious to emphasise that their New Towns marked a real break from the earlier Garden City tradition, increasing the problems for the film-makers by restricting the use of an obvious source of credible images. Initially, there was some recourse to these precedents. *Planned Town*, a film about Welwyn Garden City, officially released in 1948 but almost certainly produced earlier, was pressed into service for promotional purposes. It retains the style of earlier planning documentaries and confirms much of what was officially disliked about the Garden Cities. Its self-satisfied, middle class tone and unimaginative pedagogic style encapsulates all that needed to be avoided.

New Town (1948) attempted a quite different approach. Commissioned by the Central Office of Information from the cartoon-makers John Halas and Joy Batchelor, this ten-minute 'short' uses cartoon format to explain the basic ideas behind the New Towns. The central character, 'Charley', explains the problems of existing towns and cities and shows how these are avoided in the plans of the New Towns. In a humorous and surreal manner, the audience sees Charley and his friends shunting socially balanced housing, factories and civic buildings into place. Whether or not the end-product was more successful than the standard approaches in conveying the essence of town planning cannot now be ascertained. What is certain is that, in this approach, the process of planning was equated to a form of conjuring trick in which the elements of the town were magically rearranged into the better world of the New Town.

The apotheosis of this approach is found in *Home of Your Own* (1951). Commissioned by the Development Corporation of the New Town at Hemel Hempstead, this 22-minute film employs a human interest story to sell the New Town to possible residents. This concerns a fictional bricklayer, George Wilson, his wife Jenny and their aspirations for a better way of life that are met when they move from inner city housing in Willesden (London) to Hemel Hempstead. The film follows their progress from George's chance encounter with Hemel Hempstead while on a coach outing, through his application for employment and housing there, to the family's successful move. No complex explanations of the planning process were placed into the actors' mouths. Their concerns are with better housing, which we see first under construction and later completed.

The attempt to communicate something of the complex nature of planning is provided by a three-minute sequence in the middle of the film. The story abruptly switches from George Wilson's laborious filling-out of the application form, with his lips silently enunciating the words as he reads, to the different world on which the family's dreams depend. The audience is admitted to a meeting of the Development Corporation, presided over by Lord Reith. The action makes immediate connection with the Wilsons' circumstances as the Housing Manager reports on selection procedures for residents. Bricklayers are specifically mentioned as being in demand. After that, the meeting moves on to the next item on the agenda. This is deliberately chosen to be as abstruse as possible, namely seeking 'authority to go ahead with the first surplus water outfall culvert contract and the conversion of the King's Langley gravel pit into a balancing tank for surface water'.

As the Chief Engineer begins his lengthy explanation pointing to a complex wall-map, the scene changes to a fishing float bobbing on the canal. His voice gradually fades out and the window on this separate world of planning closes. A narrator with a polished middle-class Home Counties accent seems to take pity on the audience, clearly unable to follow such matters: 'Yes, well that's how new towns start, with drains, water supply and so forth. Things that sound unromantic and even dull but like anything that's built to last, it's the foundations that count.' After the camera cuts to pictures of an excavator digging trenches for storm drains, he continues: 'They're the basic things that the planner has to worry about. He doesn't plan for the sake of planning. His job is getting the best from what is available to him.'

The remaining part of this sequence proceeds to inform the audience, in the most general terms, how planners work to guarantee that everything is in its proper place, how they would ensure that Hemel Hempstead remained a country town and how planners strive 'to produce a better way of living'. Then, almost as abruptly as this sequence began, the audience is returned to the tale of the Wilsons and the accents of working-class Londoners. Viewed thematically, therefore, the film suggests that while ordinary people can grasp housing issues, planning is a technical and remote activity beyond their understanding. Its alchemy is best left 'to the experts'.

PORTRAYING THE PLANNER

Just as planning proposals were difficult to represent in an engaging manner, so too was the character of the planner. Since the 1930s, and especially in the war years, a new professional practice of planning had been emerging. It drew heavily on principles first enunciated before 1914 by Patrick Geddes (e.g. see Geddes, 1915). The approach placed absolute reliance on very extensive survey work, especially of social and economic conditions that had been largely overlooked in earlier plan preparation. This was followed by detailed analysis leading, by an inductive logic that was never explicitly articulated, to the actual plan.

For the practising planner this approach had two somewhat contradictory consequences. The need for surveys, extending well beyond traditional concerns only with topographic conditions, placed a new priority on teamwork, bringing together different sets of expertise. At the same time the somewhat mysterious step from analysis to actual planning proposals allowed immense significance to be attached to the 'master planner' who presides over the entire process. It was he (and it always was a 'he') who could therefore be understood as contributing the inspirational dash of genius that would finally give brilliant expression to the *mélange* of collective needs and wants that the surveys and analysis had uncovered. Invariably it was the master planner's name that appeared on the plan, even though the scale of the endeavour manifestly took it beyond the compass of one person, no matter how gifted.

Cinematically, then, the shifting nature of planning practice offered rich potential. Many film-makers did not explore this potential, presumably reasoning that it was what planning *offered* rather than how it was undertaken that had to be the most important element in what were usually quite short films. For the most part, therefore, planners are shown as anonymous figures earnestly inspecting and recording the problems of the city or scurrying around drawing offices, preparing sophisticated maps and charts. They are generally represented as anonymous technicians performing complex tasks for the greater good. Yet for all this the outcomes of their work were supposed to be readily comprehensible to a mass audience, something which inevitably brought heroic simplifications, as for example when the planner in *When We Build Again* literally rubs out and replans a whole slum area of Birmingham in little over a minute. In this and many other cases the planner's anonymity is reinforced by an absence of synchronised sound and the use of an off-camera narrator.

Despite this, there were various occasions on which planners were allowed to speak directly, and usually rather self-consciously, to camera in accents strongly suggesting middle-class upbringing and education. To get some idea of the strategies used, it is worth considering the varying treatment given in documentary film to Sir Patrick Abercrombie – one of the few figures who made a prominent appearance in more than one film. In large measure Abercrombie's cinematic prominence accurately reflected his importance within the planning profession. He was *the* 'master planner' of the 1940s.

In *Proud City*, Abercrombie was introduced as the 'world famous British authority on town planning', though the circumstances of the plan's creation obliged him to share the limelight. As noted on p. 68, he appears on camera in the company of the London County Council's Chief Architect, J. H. Forshaw. As co-authors of the *County of London Plan* (Forshaw and Abercrombie, 1943), they take turns to deliver thoughts about the broad principles that had contributed to the ensuing plan. As such, the film-maker presents them in the way that reflected the typical dual roles of the 1940s' planner. First, they appear as gifted individuals capable of solving real problems, revealed by extensive empirical survey. Second, they are presented as team-leaders, deferring on the detailed aspects to other team members. While they provide prefatory remarks about the Plan, it is left to one of their junior colleagues (Arthur Ling) to fill in detail about its content and implementation. This image of collaboration on a shared project is one that would have been much valued in the mid-1940s, with the echoes of wartime teamwork and achievement.

It also had the added benefit of giving prominence to someone who was more comfortable in front of the camera and whose accent was perceptibly more 'classless' than those of his superiors. Neither Abercrombie nor Forshaw appears at ease before the camera or well suited to the job that he had been allocated by the film-makers. Forshaw, seated at a desk, speaks his lines in a wide-eyed, slightly petrified manner. Abercrombie is hampered by his patrician accent and strained facial expression induced by wearing a monocle. He is also uncertain what he was supposed to do when Forshaw was speaking. Sitting on the desk next to Forshaw at the start of this sequence, he then wanders aimlessly round the room when not speaking. In one scene Abercrombie is seen to turn his back on the camera, pull out a handkerchief, and meander over to the mantelpiece where he proceeds to polish his monocle. Forshaw meanwhile is making comments about the social purpose of the city. Although probably inadvert, the impression of Abercrombie up-staging his colleague is unavoidable.

Perhaps learning from this experience, Jill Craigie's *The Way We Live* depicts Abercrombie in a quite different manner that exploited to the full the potential of the 'master planner' role. Here Abercrombie was clothed in the guise of hero and genius, a traditional 'architectural fiction' (Saint, 1983: 1) which the planning profession had eagerly appropriated as it established its own professional legitimacy.[16] He was presented as a visionary; an artist who takes his inspiration from the spirit of the places in which he worked and a gifted individual who could see beyond the shortcomings of today to reveal a better tomorrow. At the start of the film, indeed before the credit titles, the narrator comments that:

> The heroes or villains [of this film], according to your point of view, are two men with a plan: James Paton Watson, the City Engineer, and Professor Patrick Abercrombie. What they have to say is something of a challenge to the way we live.

The main portrayals of Abercrombie as the man of vision are found in two major sequences in the film. In the first, Abercrombie is depicted in the guise of the mystic, standing on higher ground and gazing silently at the devastation around him. The narrator notes: 'No one knew what the Professor was up to.' As he walks around the devastated centre of Plymouth, hands clasped behind his back, Abercrombie takes the audience into his confidence in voice-over. He apparently sees with the eye of an artist, for example, observing that: 'Plymouth needs pale colours to respond to the sunlight. Buildings in limestone and concrete. Flat and vertical masses to give balance to an interesting skyline. What is needed is a city to cheer people up.' As he wanders, he savours the atmosphere and reveals sources of inspiration for his plan, confiding that there could be a cardinal axis here and a precinct there. To support this theme, panoramas of existing city scenes dissolve into glowing artists' impressions of how the scene might look if the plan were implemented.

Considered dispassionately, this is all somewhat unlikely. Abercrombie was involved in many city planning exercises in the early 1940s and the ideas expressed for Plymouth fit in with the pattern-book of design ideas that were employed in those cities (Dix, 1981). It is more likely that large elements of the plan were general ideas adapted to fit the needs of the site and that Abercrombie's re-enacted tour was simply a cinematic device.

The second sequence showed Abercrombie as the impassioned visionary, willing and able to take on critics when his ideas are attacked. Paton Watson and Abercrombie are seen addressing a public meeting. At first, the audience appears restless while they watch animated analyses of Plymouth's housing and traffic problems. Several times, Abercrombie is forced to respond vigorously to interjections from the floor. Nevertheless as he turns to the plan itself and to illustrations of the new city centre, the audience becomes progressively more engrossed in his presentation. After the final illustration – a sweeping vista in honour of fallen servicemen – the audience murmurs its approval, ending with a round of applause and a dramatic crescendo in the film's music score. Doubters are still heard and some of the audience remain confused, but the visionary emerges unbowed.

CONCLUSION AND AFTERWORD

These differing portrayals of Abercrombie as scientist or artist, team-leader or solitary genius, once more point to the central problem faced by documentarists. Documentary films of the period under consideration here strove for delicate balance between instruction, entertainment and evangelism for change. Planning was regarded as an important activity, vital to the reconstruction of the cities and the regeneration of Britain. Yet it was repeatedly found difficult to convey that point in a way that would capture the imagination of a lay audience. Despite their best endeavours, film-makers found their subject matter remained obstinately remote from their audiences. While it was reasonably easy to convey a sense of urgency and activity by imaginative use of montage and voice-overs, the substance of planning was difficult to encapsulate by low-budget graphics and the woodenly intoned utterances of practising planners.

In the best tradition of a film-making genre that sought to supply a window on the world rather than passively holding up a mirror to reality, documentarists began using greater amounts of human interest to convey their story-lines. The depiction of planners gradually took second place to those stories. Similarly, there was greater use of acted sequences. At first, these were adjuncts to the main story as in *When We Build Again* but later, as in *Home of Your Own*, the acted sequences became the main vehicle for the film's message.

To some extent, the changes in presentation reflected the fact that, over the years between 1935 and 1952, the message itself had changed. The housing documentaries of the 1930s had made a dramatic case for intervention in urban affairs, but growing realisation that housing reform needed to be set in context led the film-maker in the late 1930s and early 1940s to argue the case for planning. In making that case, they took advantage of various existing notions to represent town planning as a rational application of science, as social medicine and as revelation. None was to prove entirely satisfactory. By default, a fourth strategy arose. Town planning could now be likened to wizardry, able to work profound magic for society but beyond the grasp of ordinary people.

With the general acceptance of the case for planning, symbolised by the passage of the 1947 Town and Country Planning Act (see Ward, 1994, especially 106–12), the need to press the case for planning no longer seemed so acute. To film-makers it was certainly a less exciting challenge now that the war and the main political battles had been won (see Mortimer, 1983: 114–5). To officials and ministers the prospect of yet more government-sponsored films airily promising social benefits through planning were increasingly seen as tactless and politically dangerous at a time when austerity was postponing even the most simple wants of most of the population. Discussing a prospective film in April 1947, the Public Relations Officer of the Ministry of Town and Country Planning felt it wise to give the remarkable advice to the Central Office of Information that 'the words "plan" and "planning" be kept out of the film altogether if possible'.[17] Housing, which was much closer to popular demands, was again the major focus. This tactical retreat was fully reflected in *Home of Your Own*, chronologically the last film to be considered here. In so far as it was shown at all, town planning was now depicted as a remote technical activity only practised by initiates. Accordingly the available newsreel coverage of the 1950s shows town planning drifting into the background, powerful yet unscrutinised, while the housing drive took centre stage.

The 1960s brought a revived political interest in town planning, particularly in relation to the growing problem of the motor car, and a more general concern to strengthen public participation in the planning process. These changes stimulated something of a revival in the use of film and other mass visual media for planning purposes, often at local level. By this stage, however, it was more common to employ professional communicators, particularly well-known TV journalists, to act as intermediaries between planners and public. An early example of this approach was *Traffic in Towns* (1963), where a chain-smoking James Cameron anchored what would otherwise be a highly complex technical exposition with the plain person's interpretations. It was a format that was to be adopted by many big provincial cities in the later 1960s and 1970s, usually relying on the familiar faces or voices of regional TV journalists. Furthermore, the trend towards relying on professional communicators and public relations experts was further strengthened as the priorities of planning shifted towards economic promotion in the late 1970s and 1980s. While all this lies beyond our original remit, it underlines yet again the extent to which the content of the documentary film serves as a sensitive guide to wider debates about town planning and the urban environment.

NOTES

1   In the previous year, for example, John Baxter's film *The Common Touch* (1941) had borne the opening caption: 'This picture is dedicated to the humble people of our great cities whose courage and endurance have gained for us all the admiration and support of the free countries of the world' (quoted in Aldgate and Richards: 1986: 14).

2   The closure in January 1952 of the Crown Film Unit is generally taken to mark the severance of the last direct link with the government agencies that had sponsored documentary film-making in the first place. After this date, the British documentary film changed considerably in character.

3   See, for example, Sutcliffe (1984), Minden (1985), Natter (1994) and McArthur (this volume).

4   Glasgow Corporation, for instance, commissioned a film from Gaumont in 1922 to focus on its parks and health facilities. The Health Department of the south London Borough of Bermondsey, by contrast, made a series of short films for showing to local audiences as part of a propaganda drive to promote preventive medicine (Lebas, 1992).

5   Although some would argue that these films constitute at least a distinct sub-type of the documentary genre, it is more constructive to view the difference between this type of film and the apparently less militant, more orthodox type of documentary as one of degree rather than kind (Williams, 1980: 7). In their recent book Thompson and Bordwell (1994), for example, place the documentary tradition alongside Leftist and experimental cinema for purposes of discussion.

6   In a review published in 1926, Grierson had spoken of Robert Flaherty's film *Moana*, 'a poetic vision of Polynesian tribal life', as having 'documentary value': see Hood (1983: 100).

7   Grierson's film *Drifters*, released in 1929, is generally taken to be the founding work of the documentary movement. For more on Grierson's career, see Sussex 1975), Hogenkamp (1989) and Sinyard (1989). For thoughts about the broader question of the need for more systematic research on documentary film, especially to conceptualise the problems that it raises, see Linton (1992).

8   See also the comments above in connection with the film *New Towns for Old*.

9   Although widely considered to be in the pocket of the Conservative Party (Low, 1979: 176), British Movietone News had shown an interview with Arthur Greenwood, Minister of Housing in April 1930. Greenwood argued that it was important for 'good men and women' to support the slum clearance drive. The slogan should be: 'slums must go'.

10   In passing, it is worth noting that architecture could also provide the subject material for powerful films. Ayn Rand's novel *The Fountainhead* (1943), with its plot about the modern architect as frustrated genius, led to a notable film (1949) starring Gary Cooper and Patricia Neal: see McArthur (this volume).

11 These included the containment of cities by limiting their physical size, principles of neighbourhood planning, land-use zoning, and the creation of satellite towns to receive overspill (see Ward, 1994: 80-115).

12 It is not possible to deal in detail here with the broader context of Mumford's work or the intellectual debts that he owed to Patrick Geddes (e.g. 1915). For more on Mumford, see: Hughes and Hughes (1990), Carrithers (1991) and Casillo (1992).

13 Greenbelt (Maryland), Greendale (Wisconsin), Greenhills (Ohio), Los Angeles and Long Beach.

14 This scene was misattributed elsewhere to the film *Town and Country*

Planning (Gold and Ward, 1994: 229). We are grateful to Toby Haggith for this information.

15 For more on this subject, see Hall (1988) and Ward (1994, 1995).

16 This was done in quite specific ways. For example, the emblem on the chain of office of the President of the Town Planning Institute, adopted in the 1920s shortly before Abercrombie himself became President, is a representation of 'the genius of the city'.

17 Public Record Office, file HLG 90/35, New Towns Publicity: Films: *An Englishman's Home*, Mossbacher-Forman, 16 April 1947.

FILMS CITED

The City (d. W. van Dyke and R. Steiner, 1939a). Civic Films for the American Institute of Planners.

The City (p. A. Cavalcanti, 1939b). GPO Film Unit for Anglo-American.

The Common Touch (d. J. Baxter, 1941). British National.

The Crowd (d. K. Vidor, 1928). Metro Goldwyn Mayer.

Drifters (d. J. Grierson, 1929). Empire Marketing Board/New Era.

The Fountainhead (d. K. Vidor, 1949). Warner Brothers.

The Great Crusade: the story of a million homes (p. F. Watts, 1937). Pathé Films for the Ministry of Health.

Home of Your Own (d. T. Thompson, 1951). Data Film Unit for Hemel Hempstead Development Corporation.

Housing Problems (d. A. Elton and E. Anstey, 1935). British Commercial Gas Association.

Housing Progress (p. M. Nathan, 1938). Housing Centre.

Kensal House (d. F. Sainsbury, 1937). Gas, Light and Coke Company.

Land of Promise (d. P. Rotha, 1946). Films of Fact Ltd.

Moana (d. R. Flaherty, 1926).

New Town (1948). Halas and Batchelor Cartoon Film Production for Central Office of Information.

New Towns for Old (d. J. Eldridge, 1942). Strand Film Productions for Ministry of Information.

New Worlds for Old (d. P. Rotha, 1938). Realist Film Unit for the Gas Industry.

Our Daily Bread (d. K. Vidor, 1934). Viking Productions/United Artists.

Planned Town (1948). Welwyn Garden City Company.

The Plow that broke the Plains (d. P. Lorentz, 1936). US Government Resettlement Administration.

Proud City (d. R. Keene, 1945). Greenpark Productions in association with the Film Producers' Guild for the Ministry of Information.

The River (d. P. Lorentz, 1937). US Farm Security Administration.

The Smoke Menace (d. J. Grierson, 1937). Realist Film Unit for British Commercial Gas Association.

Town and Country Planning (1946). Army Bureau of Current Affairs Magazine Series, Army Kinematograph Services for the War Office.

Traffic in Towns (1963). Realist Film Unit for Central Office of Information.

The Way We Live (d. J. Craigie, 1946). Two Cities for Rank.

When We Build Again (d. R. Bond, 1943). Strand Films for Cadbury Brothers.

REFERENCES

Aitken, I. (1990) *Film and Reform: John Grierson and the Documentary Film Movement*, London: Routledge.

Aldgate, A. and Richards, J. (1986) *Britain Can Take It: The British Cinema in the Second World War*, Oxford: Basil Blackwell.

Alexander, W. (1981) *Film on the Left: American Documentary Film from 1931 to 1942*, Princeton, NJ: Princeton University Press.

Armes, R. (1978) *A Critical History of the British Cinema*, New York: Oxford University Press.

Barsam, R. (1989) 'John Grierson: his significance today', in W. de Greef and W. Hesling (eds) *Image, Reality, Spectator: Essays on Documentary Film and Television*, Leuven: Acco, 8–16.

Bordwell, D. and Thompson, K. (1993) *Film Art: An Introduction*, fourth edition, New York: McGraw Hill.

Bressey, C. and Lutyens, E. (1937) *Highway Development Survey (Greater London)*, London: HMSO.

Carey, J. (1992) *The Intellectuals and the Masses: Pride and Prejudice among the Literary Intelligentsia, 1880–1939*, London: Faber and Faber.

Carrithers, G. H., jun. (1991) *Mumford, Tate, Eiseley: Watchers in the Night*, New Orleans: Louisiana University Press.

Casillo, R. (1992) 'Lewis Mumford and the organicist concept in social thought', *Journal of the History of Ideas* 53, 91–116.

Dix, G. (1981) 'Patrick Abercrombie, 1879–1957', in G. E. Cherry (ed.) *Pioneers in British Planning*, London: Architectural Press, 103–30.

Fielding, R. (1978) *The March of Time, 1933–51*, New York: Oxford University Press.

Ford, L. (1994) 'Sunshine and Shadow: lighting and colour in the depiction of cities on film', in S. C. Aitken and L. E. Zonn (eds) *Place, Power, Situation and Spectacle: A Geography of Film*, Lanham, Md: Rowman and Littlefield, 119–36.

Forshaw, J. H. and Abercrombie, P. (1943) *The County of London Plan*, London: Macmillan.

Garside, P. (1988) '"Unhealthy areas": town planning, eugenics and the slums, 1890–1945', *Planning Perspectives* 3, 24–46.

Geddes, P. (1915) *Cities in Evolution: An Introduction to the Town Planning Movement and the Study of Civics*, London: Williams and Norgate.

Gold, J. R. (1984) *The City in Film*, Architecture Series 1218, Monticello, Ill: Vance Bibliographies.

Gold, J. R. (1985) 'From "Metropolis" to "The City": film visions of the future city, 1919–39', in J. A. Burgess and J. R. Gold (eds) *Geography, the Media and Popular Culture*, London: Croom Helm, 123–43.

Gold, J. R. (1995) 'The MARS Plans for London, 1933–1942: plurality and experimentation in the city plans of the early British Modern Movement', *Town Planning Review* 66, 243–67.

Gold, J. R. and Ward, S. V. (1994) '"We're going to do it right this time": cinematic representations of urban planning and the British New Towns, 1939–1951', in S. C. Aitken and L. E. Zonn (eds) *Place, Power, Situation, and Spectacle: A Geography of Film*, Lanham, Md: Rowman and Littlefield, 229–58.

Greenwood, W. (1933) *Love on the Dole: A Tale of the Two Cities*, London: Jonathan Cape.

Gruffudd, P. (1995) '"A crusade against consumption": environment, health and social reform in Wales, 1900–1939', *Journal of Historical Geography* 21, 39–54.

Hall, P. (1988) *Cities of Tomorrow: An Intellectual History of Planning and Design in the Twentieth Century*, Oxford: Basil Blackwell.

Hogenkamp, B. (1989) 'The British documentary film movement in perspective', in W. de Greef and W. Hesling (eds) *Image, Reality, Spectator: Essays on Documentary Film and Television*, Leuven: Acco, 17–32.

Hood, S. (1983) 'John Grierson and the documentary film movement', in J. Curran and V. Porter (eds) *British Cinema History*, London: Weidenfeld and Nicolson, 99–112.

Hughes, T. P. and Hughes, A. C. (1990) *Lewis Mumford: Public Intellectual*, New York: Oxford University Press.

Lebas, E. (1992) 'The Council's picture show: the film-making activities of Bermondsey Health Department, 1925–48', paper presented to the London Group of Historical Geographers, Institute of Historical Research, London.

Lees, A. (1985) *Cities Perceived: Urban Society in European and American Thought, 1820–1940*, Manchester: Manchester University Press.

Levin, G. R. (1971) *Documentary Explorations: 15 Interviews with Film-Makers*, Garden City, NY: Doubleday.

Linton, J. M. (1992) 'Documentary film research's unrealised potential in the communication field', *Communication* 13, 85–93.

Low, R. (1979) *Films of Comment and Persuasion of the 1930s*, London: George Allen and Unwin.

Macfarlane, J. (1987) *Catalogue of Films and Television Programmes on Architecture, Town Planning and the Environment*, London: the author.

Marris, P. (1983) 'Notes' to accompany *Britain Can Make It?: A Film Portrait of Britain*, Newcastle-upon-Tyne: Tyneside Film Festival.

Marwick, A. (1974) 'The 1930s: the Film Evidence', A305 'Art and Design, 1890–1939', Open University televised programme, BBC2.

Minden, M. (1985) 'The city in early cinema: *Metropolis, Berlin* and *October*', in E. Timms and D. Kelley (eds) *Unreal City: Urban Experience in Modern European Literature and Art*, Manchester: Manchester University Press, 193–213.

Mortimer, J. (1983) *Clinging to the Wreckage: A Part of Life*, Harmondsworth: Penguin.

Mumford, L. (1938) *The Culture of Cities*, London: Secker and Warburg.

Natter, W. (1994) 'The city as cinematic space: modernism and place in *Berlin, Symphony of a City*', in S. C. Aitken and L. E. Zonn (eds) *Place, Power, Situation, and Spectacle: A Geography of Film*, Lanham, Md: Rowman and Littlefield, 203–27.

Paton Watson, J. and Abercrombie, P. (1943) *A Plan for Plymouth: The Report Prepared for the City Council*, Plymouth: Underhill.

Pronay, N. (1982) 'The political censorship of films in Britain between the wars', in N. Pronay and D. W. Spring (eds) *Propaganda, Politics and Film 1918–45*, London: Macmillan, 98–125.

Rand, A. (1943) *The Fountainhead*, New York: Dutton.

Richards, J. (1984) *The Age of the Dream Palace: Cinema and Society in Britain, 1930–1939*, London: Routledge and Kegan Paul.

Saint, A. (1983) *The Image of the Architect*, New Haven, Conn.: Yale University Press.

Sinyard, N. (1989) 'Grierson and the documentary film', in B. Ford (ed.) *The Cambridge Guide to the Arts in Britain*, volume 2 'The Edwardian Age and the Inter-War Years', Cambridge: Cambridge University Press, 246–53.

Snyder, R. L. (1993) *Pare Lorentz and the Documentary Film*, Reno: University of Nevada Press. (Originally published 1968, Norman: University of Oklahoma Press.)

Solomon, S. J. (1970) 'Film as an urban art', *Carnegie Review* 22(4), 11–19.

Sorlin, P. (1991) *European Cinemas, European Societies, 1939–1990*, London: Routledge.

Sussex, E. (1975) *The Rise and Fall of the British Documentary: The Story of the Movement Founded by John Grierson*, Berkeley: University of California Press.

Sutcliffe, A. (1984) 'The metropolis in the cinema', in A. Sutcliffe (ed.) *Metropolis, 1890–1940*, London: Mansell, 147–71.

Thompson, K. and Bordwell, D. (1994) *Film History: An Introduction*, New York: McGraw Hill.

Ward, S. V. (1994) *Planning and Urban Change*, London: Paul Chapman.

Ward, S. V. (1995) 'The politics of the New Towns movement', *New Town Record*, Manchester: Planning Exchange (CD-ROM).

Webb, M. (1987) 'The city in film', *Design Quarterly* 136, 3–34.

Williams, C. (ed.) (1980) *Realism and the Cinema: A Reader*, London: Routledge and Kegan Paul.

Wilson, A. (1992) *The Culture of Nature: North American Landscape from Disney to the Exxon Valdez*, Oxford: Blackwell.

Wiseman, N. (ed.) (1979) *The Image of the City in the Cinema*, Toronto: York University Press.

4

SOMETHING MORE THAN NIGHT
tales of the *noir* city
●

Frank Krutnik

The *noir* city of Hollywood's thrillers of the 1940s and early 1950s is a shadow realm of crime and dislocation in which benighted individuals do battle with implacable threats and temptations.[1] Often a little too conveniently framed as a symptomatic response to the cultural and social upheavals besetting the US after the Second World War – the nuclear age, the Cold War and homefront anti-communism, the adjustment to a postwar economic order – *film noir*'s resonant scenarios of fear, persecution, and disjunction actually began to appear before the US entered the war. *Film noir* also inherited many of its narrative and stylistic features, and much of its urban atmosphere, from the hard-boiled pulp fiction of the interwar period (Krutnik, 1991 33–44). Looking back from the vantage-point of 1950, Raymond Chandler suggested that the postwar climate was responsible for feeding, not breeding, the 'smell of fear' generated by the pulp crime stories:

> Their characters lived in a world gone wrong, a world in which, long before the atom bomb, civilization had created the machinery for its own destruc-tion and was learning to use it with all the moronic delight of a gangster trying out his first machine-gun. The law was something to be manipulated for profit and power. The streets were dark with something more than night
> (Chandler, 1973b: 7)

These 'hard-boiled chronicles of mean streets' (Chandler 1973a: 196) were first published in *Black Mask* magazine, shortly after the 1920 census revealed that over 50 per cent of the American population now lived in cities.[2] Howard Chudacoff credits this finding with symbolic importance, implying that the city had supplanted the farm as 'the locus of national experience' (Chudacoff, 1975: 179).

From Thomas Jefferson on, the city has frequently been decried as a vortex of corruption, a world fallen from the state of grace that blessed an earlier imagined America in which the small-scale agrarian community was the bedrock of demo-cratic order.[3] This restrictive Jeffersonian fantasy of the US was challenged progres-sively through the latter half of the nineteenth century, as the country was rapidly transformed into a mature industrial nation (Muller, 1993: 72). As the showcase of the industrial age, the city insisted on a definition of the American experience that threatened the hegemony of middle-American populism; for example, as a legacy of the mass immigration that fuelled expansion, the boom cities of the Northeast and Midwest were complex, multi-ethnic social environments (Muller, 1993: 71). Nevertheless, the Jeffersonian mystique (Rourke, 1973: 439) persisted in the rhetoric of populist thinkers such as philosopher Thomas Dewey, who cautioned in 1927 that 'democracy must begin at home, and its home is the neighborly community' (quoted in White and White, 1973: 424).

As Chandler saw it in his 1944 essay 'The Simple Art of Murder', the pulp crime writers were chroniclers of a modern America dominated by the concept of the city. They dealt, he wrote, with:

> a world in which gangsters can rule nations and almost rule cities, in which hotels and apartment houses and celebrated restaurants are owned by men who made their money out of brothels, in which a screen star can be the finger man for a mob, and the nice man down the hall is a boss of the numbers racket; a world where a judge with a cellar full of bootleg liquor can send a man to jail for having a pint in his pocket, where the mayor of your town may have condoned murder as an instrument of money-making, where no man can walk down a dark street in safety because law and order are things we talk about but refrain from practising; a world where you may witness a hold-up in broad daylight and see who did it, but you will fade quickly back into the crowd rather than tell anyone, because the hold-up men may have friends with long guns, or the police may not like your testimony, and in any case the shyster for the defence will be allowed to abuse and vilify you in open court, before a jury of selected morons, without any but the most perfunctory interference from a political judge.
>
> (Chandler, 1973a: 197)

However, Chandler's resonant collage speaks of the vitality of the *noir* city as well as its appalling corruption, of its enticements as well as its horrors – a dichotomy that is shared by the dark city of Hollywood *noir*. Because *film noir* communicates most expressively through its silences, evasions and disavowals, its vision of the *noir* city resists easy mapping,[4] so I will begin with a film that offers a far more clear-cut reading of the benighted modern city grounded in a populist Americanism that is rendered incredible in the *noir* thrillers themselves.

ABYSMAL CITY

It's A Wonderful Life was released in 1946, when the Second World War was just over and the cycle of *film noir* thrillers was in full swing. An unusual hybrid of supernatural fantasy, drama and comedy, the film marked the return to civilian

4.1   George Bailey (James Stuart) in the Bedford Falls of Capra's *Its a Wonderful Life*
© Republic Entertainment Inc. Courtesy of BFI Stills, Posters and Designs.

film-making of Frank Capra, a director renowned for his prewar series of social comedy–dramas.[5] *It's A Wonderful Life* seeks to reaffirm the faith of a man who sacrifices his dreams of travel and adventure to serve the needs of his small-town community, Bedford Falls (Plate 4.1). Introduced at a point of crisis, George Bailey (James Stewart) prepares to commit suicide when his family's building and loan business is on the verge of collapse. He is saved, though, by the intervention of a supernatural agent, trainee guardian angel Clarence (Henry Travers) who leaps from the bridge just ahead of George, thus transforming the attempt to take a life into a bid to save one. To persuade George that his personal sacrifices have been worthwhile, Clarence then shows him what might have become of Bedford Falls had he never been born. He presents him with a nightmare vision of modern America – a vision of the *noir* city (Wood, 1986: 65).

In this alternative reality, the idealised middle-American community of Bedford Falls is supplanted by the wild urban world of Pottersville (Plate 4.2). As George runs through the town, a series of tracking shots exposes the iconography of the 1940s *noir* city: a riot of neon and jazz, the main street of Pottersville is crammed with burlesque halls, dance joints, pawnbroker's stores, numerous bars. This unholy

4.2   Bedford Falls becomes Pottersville: the 'unborn sequence' of *Its a Wonderful Life*

place is the realization of George's nemesis, Old Man Potter (Lionel Barrymore), a greedy, wheelchair-bound capitalist. George is a bulwark against Potter's influence: through the building and loan company, inherited from his father, he maintains Bedford Falls as a folk community of family, friends and neighbours by lending to the 'common people' so they can build homes. With George Bailey, himself a family man and man of the people, capitalist enterprise helps to service the community – it is essentially, fantastically procreative (after the Building & Loan survives a financial crisis during the Depression, George places the company's two remaining dollar bills in the vault, and christens them 'mama dollar and papa dollar'; *Framework*, 1976: 22). Pottersville's urban world, however, is a wasteland of non-productive excess, its public spaces dominated not by family homes but by the brash seductions of commercialized leisure. This spectacle of urban degeneracy signifies a corruption of the social body, which the film links to the 'crippled' body of Potter himself. Where George builds himself a family, and helps others to do the same, Potter's desire for money is a 'disabled' accumulation, without issue. Where George builds for life, Potter's capitalist excesses are in the service of the death drive – Bailey Park, the family housing estate of Bedford Falls, is replaced by the cemetery, 'Potter's Field'. In this barren double of Bedford Falls, George's family is especially blighted – his wife, Mary (Donna Reed) is now a neurotic spinster librarian, and his mother (Beulah Bondi) administers not to her own family but to a household of strangers in a down-at-heels boarding house.

Like the protagonists of many *noir* thrillers, George stumbles through the streets of Pottersville in search of the refuge of the familiar – only to find his intimates mutated into hardened strangers, and the community ethic of Bedford Falls eroded by self-protectionism and fear. The moral, Clarence tells George, is that 'Each man's life touches so many other lives – and when he's not around, he leaves an awful hole, doesn't he?' But the horror of Pottersville comes not simply from the transformation of those George knows and loves, but from the dislocation of his own identity. The Pottersville nightmare – tagged the 'unborn sequence' during the production (McBride, 1992: 520) – is predicated upon George's absence, yet he is forced to *inhabit* the uncanny landscape of the familiar made strange.[6] Roaming the benighted streets – deprived of his history and the meaningful coordinates of being – he is radically 'homeless'.[7] The conflict between desire and responsibility that plagues George is finally purged through his descent into Pottersville – which is a black hole, a vortex of negation that strips away and reveals the sustaining principles of his life. George is immediately rewarded for his restoration of faith, his *rapprochement* with the familiar, as the people of Bedford Falls descend upon his home, donating funds that will save the Building & Loan and keep him from prison. It is a soaringly sentimental resolution, in which the ordinary folk of Bedford Falls reciprocate George's earlier sacrifices, to suture him back into the community after his spell in a wilderness of self-doubt. The Christmas setting for the film's conclusion adds symbolic weight to George's rebirth, sanctifying the small-town and the individual's place within it.[8]

Bedford Falls and Pottersville correspond to the two contrasting forms of social organization designated by Ferdinand Tönnies as *Gemeinschaft* and *Gesellschaft*. As Phillip Kasinitz outlines:

> For Tönnies, *Gemeinschaft* is a type [of] social solidarity based on intimate bonds of sentiment, a common sense of place (social as well as physical), and a common sense of purpose. *Gemeinschafts*, he argues, are characterized by a high degree of face-to-face interaction in a common locality among people who have generally had common experiences. The sense of social norms is strong and individual deviation is relatively rare. There is a high degree of social consensus and behaviour is governed by strong but usually informal institutions such as the family and peer group. In a *Gesellschaft*, in contrast, relationships between people tend to be impersonal, superficial and calculating, and self-interest is the prevailing motive for human action. Social solidarity is maintained by formal authority, contracts and laws.
>
> (Kasinitz, 1995: 11)

The folk community of Bedford Falls resembles Thomas Jefferson's pastoral ideal, a realm of localized Americanism protected from the pestilence of urbanity. And Pottersville is a corrupted city of strangers that has betrayed the Edenic promise of America – a world in which consensual social bonds have been obliterated under the pressures of unchecked capitalism.[9]

However, such a nostalgic fantasy of the 'eternal' American community was no viable model for mapping the complex social landscape of postwar America, and in attempting to rework the optimistic, folksy spirit of Capra's Depression-era triumphs *Mr Deeds Goes to Town* and *Mr Smith Goes to Washington*, *It's A Wonderful Life* retreats from their 'passionately engaged social criticism' (McBride, 1992: 522).[10] In his review of the film, James Agee complained that

in representing a twentieth-century American town Frank Capra uses so little
of the twentieth and idealizes so much that seems essentially nineteenth-
century, or prior anyhow to the First World War, which really ended that
century. Many small towns are, to be sure, 'backward' in that generally more
likable way, but I have never seen one so Norman Rockwellish as that.

(quoted in McBride, 1992: 522)[11]

Another telling difference from Capra's earlier work is the more restricted options
that the film accords its women. In *Mr Deeds*, *Mr Smith* and *Meet John Doe*, the
conflict between the hero and the modern city is mediated through his relationship
with an urban career-woman. Following what Charles Maland (1995: 111) has
termed a 'cynicism–conversion–faith pattern', the woman's trajectory crystallizes the
general process through which the hero restores faith and integrity to an urban
world under the sway of cynicism, exploitation and corruption. By contrast, the
central female characters of *It's A Wonderful Life* – George's mother and wife –
play more passive supporting roles,[12] and the vital male–female contest of Capra's
prewar films is replaced with a conflict that takes place *within* the male protagonist.[13]

Film noir's vision of the abysmal city flaunts ambivalences about the relationship between the individual and the community that Capra's film struggles to
contain. In *D.O.A.*, for example, Frank Bigelow (Edmond O'Brien) is lured to the
city by the promise of hedonistic escape from his small town – whose name,
Banning, leaves no doubt about its limitations. Bigelow's escape into the city of
excess (bars, jazz, available women) reaps immediate consequences: slipped a fatal
dose of uranium poison in a nightclub, he spends his final hours running through
the city streets, struggling, as he expires, to comprehend what it is about his dream
of escape that has warranted such a death. The dark enticements of the libidinal
city also contaminate the women at the centre of *Shadow of a Doubt* and *The
Reckless Moment* – provoking in each an internal struggle that exposes such a
multitude of dissatisfactions with their small-town regime that the final return
to order can only ever be an uneasy compromise. Although *film noir* locates the
modern city as a threat to the 'American community', it refuses to sanction
the small town as a redemptive alternative. Moreover, it is also more willing to
acknowledge that the city of strangers may hold attractions barred from the
restricted orbit of small-town America.

THE PRIVATE 'I'

Like George Bailey, *noir* protagonists tend, at one time or another, to find themselves with nowhere to run, nowhere to hide, nowhere to call home.[14] But where
Capra's hero has somewhere to return to, the *noir* thrillers imply the increasingly
phantasmic nature of home and community in modern America. The problematizing
of the certainty of 'home', Sylvia Harvey (1978) suggests, is signalled by the 'terrible

absence of family relations' (p. 33) in the *noir* thrillers, and by their preoccupation 'with the loss of those satisfactions normally obtained through the possession of a wife and the presence of a family' (p. 27). And Richard Dyer points out that, in *film noir*,

> settings tend to be in the public world rather than domestic. For the hero a basic domestic ritual like eating is transferred from family to public eating place. Indeed, the lunch counter comes close to being one of the true icons of the form. . . . Crucial personal encounters take place not in the home but, say, in a train (*Strangers On A Train*), in a supermarket (*Double Indemnity*) or in a seedy cafe (*Out Of The Past*). In this way the hero is denied an environment of safety, coziness, or rootedness. If such an atmosphere is evoked at all, it serves to sharpen the depiction of the *noir* world by being under threat from the latter (*Kiss Of Death*) or actually destroyed by it (*The Big Heat*). . . . More usually, when homes are shown, they are the homes of the villains and moreover are 'abnormal' – they belong to the single (ie. 'incomplete') people as in *Laura*, or childless couples as in *Gilda* or [*The*] *Postman* [*Always Rings Twice*], or, of course, gays as in *Rope*. That these homes are abnormal is iconographically expressed once again in a style of luxury quite different from the cozy normality of the 'ordinary family home'.
>
> (Dyer, 1977: 19)

The *noir* city is the principal stage upon which this drama of the obliteration of 'home' is played out – but instead of dealing directly with the social forces that have made the modern city so 'unliveable', *film noir* fixates upon the psychic manifestations of such dis-ease. In the arena of the *noir* city, protagonists must confront both the strangeness of others and the strange otherness within – as *film noir*'s scenarios of disorientation and dislocation challenge their ability to chart an identity in *noir*'s expressionistic simulacrum of modern America.[15]

The most iconic male inhabitant of the *noir* city is the private detective, who shares with the *femme fatale* a wilful rejection of domestication. The depraved, lawless city Chandler describes in 'The Simple Art of Murder' serves the private-eye as a new frontier – like the old frontier of the West, it is a readymade venue for spinning out romanticized fictions of male adventure.[16] But where the hero of the classic Western story is torn between the familial community and the wilderness, the *noir* detective is a privatized hero, customized to the atomistic regime of *noir*'s urban *Gesellschaft*. Chandler's serial detective, Phillip Marlowe, may expose the sewers of corruption beneath the ordered façade of the city, but his diligent efforts result in little change – when the case is closed, the city remains lawless. As Stephen Knight argues, in Chandler's fiction the focus upon Marlowe's private conflicts – as a man of the city but a man against the city – continually displaces the social and political causes of its disordered state (Knight, 1988: 79–85). Chandler's essay defines his detective as a privatized man of honour who resists the base seductions of the modern world:

down these mean streets a man must go who is not himself mean, who is
neither tarnished nor afraid. The detective in this kind of story must be such
a man. He is the hero, he is everything. He must be a complete man and a
common man and yet an unusual man. He must be, to use a rather weathered
phrase, a man of honour, by instinct, by inevitability, without the thought
of it, and certainly without saying it. He must be the best man in his world
and a good enough man for any world. I do not care much about his private
life; he is neither a eunuch nor a satyr; I think he might seduce a duchess
and I am quite sure he would not spoil a virgin; if he is a man of honour
in one thing, he is that in all things.

(Chandler, 1973a: 198)

As a privatized hero ('private I'), Philip Marlowe's main aim is to preserve his
integrity and independence. Knight suggests that:

Marlowe largely makes his own case to investigate: he is a private detective
in being outside the police, but he is also a *really* private eye, choosing what
he wants to look at. The positive values, as well as the negative features of
Marlowe, centre on the separate individual, who needs to be strong, active,
defended against dangers and temptations.

(Knight, 1988: 80)

The impact of the American private-eye as a culturally iconized fantasy male
derives from his role as a perpetually liminal self who can move freely among the
diverse social worlds thrown up by the city, while existing on their margins. As
Ron Goulart puts it, he 'can take you anywhere. You can go down in the ghetto,
you can go down to the underworld – but you can also go into the haunts of the
rich and famous. He gave you access to almost any level of society' (Annenberg,
1994). But at the same time as Marlowe traces the secret connections between the
city's legitimate façade and its underworld, he compulsively replays the process by
which he can remap the self as a stable entity that is separate from the degradations
of the world he lives in. Striving constantly to justify his status as private 'I', the
detective rejects the claims of social identity. Marlowe's contact with others is
fleeting, and is generally conducive to paranoia – as nearly everyone he meets tries
to bribe him with money, sex, or lies. Even his contractual bond to the client –
usually, a succession of clients – is more a means of legitimizing the process by
which Marlowe can engage with the self-defining challenges of the city. Writing in
1947, John Houseman complained of the contemporary screen incarnations of
Phillip Marlowe that 'In all history I doubt there has been a hero whose life was
so unenviable and whose aspirations had so low a ceiling' (Houseman, 1947: 162).
As the continuous interior monologue of Chandler's fiction implies, the privatization
of the self has its cost – despite the freedom and mobility Marlowe enjoys in
the city of strangers, he ultimately remains trapped within his self-containment.
To remain 'uncontaminated', Marlowe must resist emotional entanglement – and

anyone who threatens sexual temptation can provoke alarming outbursts of vituperation.[17]

In the *film noir* thrillers, the anxieties provoked by the privatized self and the voiding of community tend to suffocate the adventure – or, at least, the adventure becomes a means of trying to dispel such anxieties. Most representative of the *noir* thrillers are not the relatively controlled private-eye heroes of *The Maltese Falcon* and *The Big Sleep*, but the men at the centre of *The Dark Corner* and *Out Of The Past* – tough detectives, styled after Hammett and Chandler, who have descended into masochistic abnegation (Krutnik, 1991: 100–14). The traumatic disjunctions experienced by such *noir* protagonists connect with broader anxieties about identity, gender and the decay of community within the 'cultural unconscious' of the modern US – but they do so in a manner akin to the Freudian dreamwork, with its elaborately orchestrated condensations and displacements. *Film noir* extends the process by which Hollywood fictions codify social conflicts in individualistic terms, by grounding its logic of character and action in the protocols of popularized psychoanalysis, which had a widespread impact in the US by the 1940s (Krutnik, 1991: 45–55). Popular psychoanalytic discourse intensified the *noir* thrillers' investment in the privatized individual – encouraging the definition of the self in intra-subjective terms – while the reputation of psychoanalysis as a scientific master-discourse that could unlock the mysteries of the psyche served to legitimate the films' exploration of the murky backwaters of the unconscious.

As Pottersville is alive with shadows emanating from within George Bailey – born of his wish to be unborn – so too the mysteries and conspiracies of the *noir* city seem shaped by the projection of disordered and disorderly desires. 'The basic structure of *film noir*', Richard Dyer observes, 'is like a labyrinth with the hero as the thread running through it' (Dyer, 1977: 18). The labyrinth is formed not just from the tortuous complexity of narrative intrigues, but also from the recurrent use of representational strategies – such as voice-over, flashback, optical point-of-view, simulated dreams, nightmares or hallucinations – that serve to engulf the drama in a vortex of subjective overdetermination.[18] Those who journey through this expressionistic labyrinth of urban nightmare are blocked continually by the unmappability of the world 'out there', a world in which clear-cut boundaries between self and other have been obliterated. The spectacularly strange narrative logic of the *noir* thrillers exacerbates such problems: for example, the outrageous coincidences of *Detour*, *Dark Passage* and *They Won't Believe Me* force an uncanny elimination of the gap between desire and its object – in each case, people persecuting the hero 'just happen' to die in his presence. Al Roberts (Tom Neal), the tortured protagonist of *Detour*, ascribes such strange occurrences to the merciless whims of 'Fate, or some mysterious force', but in *film noir* Fate is the prime metaphor for the unconscious – whose 'other' logic speaks through the rupturing of credibility wrought by chance and coincidence.

Haunted by the overpresence of the unconscious, the streets of the *noir* city are often curiously empty – the public, social world overwhelmed by privatized traumas. The enclosedness and interiority that were vital to the atmosphere of

4.3   *When Strangers Marry*

Hollywood's dark city were intensified by the practice of studio filming;[19] even during the postwar vogue for location shooting, the display of 'authentic' city spaces in urban crime thrillers tended to be combined with the chiaroscuro styling of *noir* studio productions.[20] Two further institutional determinants of *film noir*'s expressionistic style are worth noting. First, Hollywood film-makers were well versed in the subterfuges of subtextualisation as a way of compensating for the absences forced by the Production Code (Hays Code) from the early 1930s. The alluring and allusive representational techniques of *film noir* – its vividly resonant atmosphere, shadowy *mise-en-scène*, and lively line in metaphorical repartee – hint continuously at a parallel universe of enticement and horror. Second, as Paul Kerr has considered, B *films noirs* – subject to limited budgets, shooting schedules and running times – developed an especially concentrated and stylized mode of story-telling that invested in non-realist techniques, such as back-projection and montage-sequences composed of stock footage (Kerr, 1978: 51–6).

This chapter concludes with a detailed examination of a remarkable B-*noir* film, the 1944 suspense thriller *When Strangers Marry* (also known as *Betrayed*; Plate

4.3).[21] Lauded for its portrayal of New York city – even though it was shot in the studio with minimal sets, cast and budget (Miller, 1987: 311) – *When Strangers Marry* attempts not to capture the 'actuality' of New York as a place, but to catch some of its resonances as the American city that most concretely and complexly symbolized the modern world.[22] Constructed, like many B-*noir* thrillers, as a succession of atmospheric set-pieces, *When Strangers Marry* weaves an astonishingly intricate web of associations to create a darkly 'poetic' vision of New York as a labyrinth of strangeness and difference. The film's principal narrative enigma concerns the murder and robbery of wealthy Philadelphian Sam Prescott, a crime for which travelling salesman Paul Baxter (Dean Jagger) is hunted through most of the film – until the killer is eventually revealed to be another salesman, Fred Graham (Robert Mitchum). But the characteristic drama of imperilled masculinity generated by this crime scenario is displaced by the film's focus upon the experiences of a woman – Millie (Kim Hunter) – as she journeys to the *noir* city, searching for Paul, the man she has recently married. In one of *film noir*'s flamboyant coincidences, she also happens to be involved with Fred. *When Strangers Marry* explores the enigma of a woman's desire in the context of a fictional cityscape that is conventionally dominated by male fantasy and anxiety; as the reviewer of the British trade journal *Kinematograph Weekly* remarked, it is 'as good a woman's picture as it is a thriller' (*Kine Weekly*, 1945: 20). Moreover, the degree to which, in different ways, its three central characters are all dislocated from the traditional promise of home, family and community whispers suggestively of wider disturbances besetting contemporary America.

CITY OF STRANGERS

In one of the the film's many pregnant encounters, New York Homicide detective Lieutenant Blake (Neil Hamilton) speaks with Fred Graham in a Turkish bath:

BLAKE   OH, THE CITY'S A NICE PLACE TO LIVE AND ALL THAT, BUT I'LL
            TAKE A HOME IN THE COUNTRY ANYTIME.
GRAHAM   WELL NOW, DEPENDS ON WHAT YOU CALL LIVING.
BLAKE   WELL, YOU CAN'T RAISE KIDS IN THE CITY.
GRAHAM   WELL, I DON'T KNOW – EVERYBODY ISN'T MARRIED.
BLAKE   EVERYONE SHOULD BE.

Fred Graham has left the country behind forever, but cannot accommodate himself to the less certain mode of being that the city demands. Despite the bravado of his words to Blake – a masculine performance in the steam-room's towelled intimacy – Graham does not embrace the city pleasures of 'living', but engages in a pitiful masquerade of the domestic ideal denied to him. He is rarely seen without his two domestic props, a pipe and a faithful dog. The desperate nature of Fred's fantasy of home is highlighted by the fact that he murders in order to have a chance of realizing it, in the hope that the $10,000 he takes from Prescott will make it possible for him to marry Millie, his small-town sweetheart.

After the steam-room conversation, Blake takes Fred back to headquarters, to show him photographs of men whom the lure of 'easy money' has transformed into killers. These are ordinary men, like Fred Graham, who are condemned not because what they want is so extraordinary but because they take an extraordinary gamble in search of commonplace gratifications. Like many of the central players of *film noir*, Fred risks all on a criminal transgression – but he lacks the go-getting boldness of Walter Neff (Fred MacMurray) and Phyllis Dietrichson (Barbara Stanwyck), the thrill-seeking iconoclasts of *Double Indemnity*. Rebels against both the family and 'legitimate' capitalist enterprise, Walter and Phyllis cold-bloodedly murder her husband for a lucrative payoff from the insurance company that Walter works for.[23] They are products of the nightmare *Gesellschaft* of modern America, in which a ruthlessly competitive self-interest has displaced allegiances to other individuals and to social institutions. Unlike *It's A Wonderful Life*, though, *Double Indemnity* – co-scripted by Raymond Chandler – does not mourn the decay of communal and familial values, and refuses simply to condemn its transgressors. Described by Charles Higham and Joel Greenberg (1968: 22) as 'a film without a single trace of pity or love', it gleefully portrays these loveless lovers playing amidst the rubble of America's social idealism.

Unable to abandon the familial regime assaulted by Walter and Phyllis, Fred remains caught between the fantasy of small-town domesticity and the merciless ambitions of modern America that are symbolized by the city. His psychic and social displacement is laid bare when he runs into Millie in New York, in the salesmen's hotel where Paul has booked a room for her. Even though she has passed him over in favour of another man, Fred loyally accompanies Millie as she waits for her mysterious husband to appear. This scene opens with a medium-shot of Fred in an armchair – pipe in hand, dog on his lap, light music playing on the radio – from which the camera dollies back, fixing on a two-shot of Fred and Millie in their separate chairs. This fleeting domestic scenario is then radically undercut by the infiltration of the vibrant presence of the city – as the pulsing light from the dance-hall across the street flashes into the background. In this moment, the *mise-en-scène* simultaneously foregrounds Fred's wish for domestication and underlines its incongruity, framing in dynamic tension the conflicting desires of these two refugees from small-town America. The film cuts to Millie's resonantly doubtful expression when Fred says he is reminded of 'old times back in Grantsville': where he yearns for the lost possibility of home, Millie has left it in pursuit of something different, something stranger.

Fred Graham is a secondary figure whose dilemma casts into shadowy relief the enigma of Millie Baxter. Just as Fred's placid exterior – passive, feminized, domesticated – conceals seething undercurrents, so Millie's virginal façade belies an attraction to 'strangeness' that others find alarming. Voyaging to join the husband she has not seen since her wedding-day, a month before, the virginal bride breathlessly relates her story to the staid middle-aged couple whose compartment she shares on the crowded train to New York. Millie speaks of meeting Paul while she worked as a waitress back home in Grantsville, Ohio, and the woman comments, condescendingly: 'Love at first sight, eh?'. 'No,' comes the earnest reply, 'I didn't even pay much attention to him. He was strange. I remember our first date – I didn't want to go, but somehow that night I just found myself waiting for him. We just walked, he didn't say much. It's hard to explain, but I felt that same strangeness.' Like the woman on the train, Blake is astonished to learn that Millie knows so little about her husband: 'You mean, you only saw this man three times and then married him? . . . You've practically married a stranger.' As a voice for law and family, he recoils from Millie's pursuit of the strange desire that leads her from country to city. While the attraction of the two salesmen to Millie's aura of uncorrupted small-town wholeness is comprehensible, her gravitation to these benighted men is something she herself finds hard to fathom. Her transgressive gamble – marriage to a travelling salesman, a disreputable transient – is a rebellion against the very domestication that Fred murders for, and her journey to the *noir* city is both an adventure of desire and an initiation into the darkness that underlines the dream of American progress.

Millie is an unusual figure for a 1940s thriller, who confuses the traditional *noir* polarisation of bad-girl and good-girl (while at the same time resisting the good–bad-girl tease of Rita Hayworth's *Gilda* or Lauren Bacall's Vivien Sternwood in *The Big Sleep*). Because of her quiet determination not to settle for the limitations of the small town, Millie is a fatal woman for Fred – but she is not a *femme noir* like Phyllis Dietrichson, Brigid O'Shaunnessy (Mary Astor) in *The Maltese Falcon*, Kitty Collins (Ava Gardner) in *The Killers*, Kathie Moffett (Jane Greer) in *Out of The Past*, or Elsa Bannister (Rita Hayworth) in *The Lady From Shanghai*. These ambitious women use their sexual allure to barter for themselves the luxury and liberty denied the American housewife. Dangerous free agents, they inspire doubts within the hero not only because they operate according to their own agendas, but also because they threaten his sexual and economic prowess – on both fronts, the *noir femme* issues to the hero a challenge to prove that he can 'satisfy' her. The *femme fatale* is never simply the cause of disorder, but she does service as the principal scapegoat for far less tangible anxieties about identity, gender and economic status that circle beneath the surface of the *noir* narratives. Her mercenary impulse, for example, becomes the prime incarnation of the dark city's hostile competitiveness.[24] Millie, though, remains centred as the narrative's desiring subject – a place generally occupied by men in these films[25] – and she lacks the *noir femme*'s obsession with money. And although she is also preserved from the fetishistic eroticism that is a trademark of *film noir*, Millie, like the *femme fatale*, is appropriated as a vehicle for projected desires. Fred not only casts her in the fantasy role of 'good wife', but the manner in which he kills Prescott – strangling him with a silk stocking bought as a present for Millie – connects the act of violence to his frustrated sexual desire.

The New York of *When Strangers Marry* is a stylized, expressionistic landscape that resonates with forces unleashed from the woman's illicit self, her desiring self.

The city is a testing-ground in which this small-town Pandora must fight for the validity of her desire to desire. As Millie waits anxiously in the darkness on her first night in the hotel-room, the lonely silence is shattered by an eruption of loud, frantic jazz. Terrorized by these alien pulsations, she creeps towards the window and pulls up the blind – and the harsh light from the dance-club's sign floods into the room. As the light flashes on and off, Millie is rendered a creature half of darkness, half of light – part angel, part woman of the shadows. She is similarly taunted with her out-of-placeness in a later scene: when she learns Paul is the man suspected of killing Prescott, Millie flees from him into the dark city – and in Central Park she is accosted by the grossly-distorted faces of the people encountered on her journey of discovery, faces that thrust out from the night to accuse Paul of murder.[26] These scenes show the influence of the contemporary cycle of persecuted-woman thrillers – popular gothic films such as *Rebecca*, *Gaslight*, *Experiment Perilous*, *My Name Is Julia Ross*, *The Secret Beyond The Door*, and *The House On Telegraph Hill*, in which a woman is subjected to psychological, sometimes physical brutalization after she marries a strange, troubled male and moves into his dark house.[27] In *When Strangers Marry*, the threatening male space is not a house, but the city itself – the *noir* city, colonized by male fantasy and anxiety, that usually offers few options for women.[28] And where the gothic thrillers represent domesticity as a regime of entrapment for the woman, Millie's union with Paul opens her to different possibilities.

When she finds Paul in the city, he is sequestered in a seedy apartment under an assumed identity. Dishevelled and unshaven, he is an abject specimen reminiscent of the strung-out masochistic men of films such as *The Killers*, *The Dark Corner*, *Detour*, *Out Of The Past* and *Nora Prentiss*. *Kinematograph Weekly* had some difficulty with the motivations of Millie's mysterious husband: 'the story does not make it clear why Paul should behave in such a suspicious manner' (*Kine Weekly*, 1945: 20).[29] Keeping Paul enigmatic is necessary to maintain suspense – neither the spectator nor Millie knows for sure that he is not the killer until very late in the film – but the unreadability of his character is more than a mere plot exigency. The problematic nature of identity is a persistent concern in the film – Paul, like Millie and Fred, is not what he appears to be.[30] However, the film does offer an implicit 'explanation' for his behaviour, by sustaining a close parallelism between Paul and Fred. As Fred sits with Millie in the hotel room, he remarks with weary determinism: 'Places are all the same. . . . You can't run away from yourself.' These sentiments are later echoed by Paul, when he observes children playing in the street: 'Kids squabbling over marbles. In 20 years' time, they'll be still be squabbling. But over money instead of marbles. People don't change much, do they?' For both men – salesmen, like Walter Neff, who labour under the yoke of commercial enterprise – the corruption of the human spirit by money is an inescapable certainty. These two men, who never meet, share the same abjection – which derives, the film implies, from the lifestyle their job forces upon them ('eating in one-arm joints, sleeping in third-rate hotels,' as Paul describes it). But where Fred gives in to the impulse to go for the quick fix, the easy kill, Paul resists – although, as he confesses to Millie, he was sorely tempted: 'You go along for years, batting your brains out for $50 a week. And that $50 seems about the most important thing in the world to you. And then suddenly you meet a guy that carries [$10,000].'

Paul smartens up after Millie finds him, and he promises to show her the city of New York. The sightseeing tour is rendered, in typical B-*noir* fashion, as a compressed montage-sequence which superimposes medium-shots of the happy couple over stock views of New York locations. Although the two sets of images are yoked together, they are not integrated: Paul and Millie never interact with the real city, but remain on their private visual plane. The stylization of the montage sequence has the effect of distancing the touristic city as a generic experience, as a series of well-worn urban views and responses. Their second foray into the city, though, reveals some of the other worlds that lie in the shadow of this urban spectacle.

When the police close in, and Paul makes his bid to escape, Millie decides to accompany him – though she has no proof of his innocence. Paul's abject state opens a window for her own empowerment, enabling Millie to exchange her unchannelled desire for 'strangeness' for the 'direction' (nurturing, redemption) of a dislocated male.[31] Furthermore, by guiding Paul through the labyrinth, Millie is able to taste the life of the outlaw: after she sidetracks the police with another man's photograph, Millie comes to take the dominant role in administering Paul's flight from the law. When she pacts herself to the fugitive male, she initiates the kind of love-on-the-run fantasy found in outlaw couple films such as *You Only Live Once*, *They Live By Night*, *Shockproof* and *Gun Crazy* (Krutnik, 1991: 213–26) – but in this instance the lovers voyage not through the backroads of America but through the strange backwaters of the city.

Paul and Millie are stranded in the middle of Harlem when his paranoia causes them to abandon the 'Share the Ride' trip they have booked for Louisville. Fleeing the visibility of the streets, they enter a basement club – only to make themselves even more glaringly prominent, as the only white faces in an all-black establishment. The principal function of the Harlem club scene is to register the 'alienation' of Paul and Millie from mainstream society – by placing them as minority figures in an all-black world, where not even the policemen are white. But it also uses race to complicate the white-nostalgia model that characterizes the discourse of community in affirmatory films such as *It's A Wonderful Life* and *My Darling Clementine*. One of the first images to mark the disordered state of Pottersville is a shot of a black musician's frenzy as he pumps out boogie-woogie piano rhythms in one of the town's heaving bars. In Ford's film, Wyatt Earp first asserts his presence in Tombstone by taming the drunken Indian Charlie, and ordering him to get out of town; he also humiliates the mixed-race saloon-girl Chihuahua (Linda Darnell) by dunking her in a horse-trough and commanding her to 'get back to the reservation'.

One of the less fortunate ironies of the term *film noir* is that these 1940s tales of the modern city deny prominence to African–Americans.[32] Although *When Strangers Marry* defines black culture through its most familiar appropriations by mainstream culture (boogie woogie, jive-dancing, boxing), the scene in Big Jim's nightspot is the closest the film comes to visualizing the possibility of a

Gemeinschaft community. The scene overturns the containment of blacks as a servile minority in US society – a situation personified in the porter who assists Millie on the train to New York – and it also casts Paul and Millie momentarily adrift from their own narrative, so they can bear witness to a self-contained social world that possesses a vitality and cohesion lacking in the America they know. Without offering the Harlem subculture as a template for the disordered society that Paul and Millie are in flight from, the film does use it as a significant barometer of that society's disjunctions – as well as allowing a tantalizing glimpse of the city as a regime that accommodates multiple cultures.

The film's on-the-run sequences also reverberate with implications of familial disintegration that evoke Blake's earlier words to Fred Graham. In the share-car, it is not just the suspicious gaze of the driver that makes Paul uneasy, but also the ceaseless screaming of a baby carried by another passenger, a lone mother. Their flight is terminated by a similar 'betrayal' by a child. When they leave Big Jim's club, Paul and Millie wander aimlessly through unfamiliar, impoverished streets, until she eventually finds a room in a shabby boarding-house. The lovers are not there long, though, before the landlady's grim young daughter – also, apparently, fatherless – reports them to the police when she sees Paul sneaking into the building. Where *It's A Wonderful Life* restricts its spectacle of familial dismemberment to the dark fantasy of Pottersville, *When Strangers Marry* produces a more disturbing and uncanny suggestion of the disordering of family, by framing the couple's expulsion to the city's 'heart of darkness' between two instances in which children are their persecutors.

THE SMELL OF THE METROPOLIS

> I did a film, *Pickup on South Street*, set in New York but I shot it all in Los Angeles and the backlot at 20th Century Fox. *The New York Times* gave it an immense review: said it was tremendous and how for the first time in a long time a Hollywood movie had captured the spirit of New York, the smell of the metropolis.
>
> (Sam Fuller, quoted in Lipsitz, 1990: 189)

Like many Hollywood directors, Sam Fuller is prone to exaggerate – Bosley Crowther's *New York Times* review actually described the film as a 'highly embroidered presentation of a slice of life in the New York underworld. Sam Fuller . . . appears to have been more concerned with firing a barrage of sensations than with telling a story to be believed' (Crowther, 1953: 38). Nevertheless, Crowther's complaint about the film's lack of narrative 'substance' misses the point – as *When Strangers Marry* illustrates, it is precisely through the triggering of sensations that *film noir* speaks most eloquently. A mode of signification that privileges connotation over the denotative, cause-and-effect logic of linear narrative, the highly-wrought *noir* aesthetic ensures that the 'meaning' of the *noir* city is not to be found in the narrative's surface details but in its shadows, in the intangibilities of tone and mood. *Film noir* is a kind of 'special effects' cinema in which stylistic performance threatens continually to overwhelm the telling of the story,[33] and in the *noir* city, where sensation overwhelms sense, characters face disordering effects for which causes cannot easily be traced.

The horror of the *noir* city is firmly tied to the pervasive indeterminacy of meaning. In the fictional universe of *film noir*, intangibility is both a stylistic strategy and a thematic obsession. Dark with something more than night, the *noir* city is a realm in which all that seemed solid melts into the shadows, and where the traumas and disjunctions experienced by individuals hint at a broader crisis of cultural self-figuration engendered by urban America. The *noir* thrillers replace the certainties of *It's A Wonderful Life* with a more nuanced, more disorganized, much bleaker vision. By shaping Pottersville as the abyss that gives meaning to Bedford Falls, Capra's film presents an ordered context for framing the disorders of modern America. But *When Strangers Marry* is typical of *film noir*, both in the way it multiplies rather than contains the signifiers of dis-ease, and in its blockage of the small-town community ideal as a balance for the dark urban world – Grantsville is never shown, and is defined exclusively through the opposing fantasies of Millie and Fred.

The suspense plot of *When Strangers Marry* is resolved when, with Millie's aid, Blake traps Fred into revealing his guilt. Millie and Paul are able to take their delayed honeymoon, and the film concludes with an epilogue that reprises and reverses her first appearance. On the crowded train from New York, the couple share their compartment with a young woman (Rhonda Fleming) who announces that she has just married. This rhyming scene signifies formal closure and also serves to register how far Millie has travelled since her journey to New York – rewarded with the husband she has been hunting throughout the film, she emerges triumphant from her trials in the city. But even though it closes with a conventional affirmation of the heterosexual couple – which ordinarily figures the social integration of the protagonist – the film provides no indication of where Millie and Paul will go once the honeymoon is over. With neither the small town nor the city presented as a liveable space, this exclusive focus upon the couple becomes but one more signifier of the evaporation of social allegiances that characterizes *film noir*'s representation of modern America. Locked in their private orbit, Millie and Paul ultimately have nowhere to go, nowhere to call home. In *When Strangers Marry*, the 'smell of the metropolis' seems to emanate from the corpse of traditionalist Americanism, later to be resurrected by Frank Capra.

NOTES

1   Not all *films noir* were exclusively concerned with the modern American city: the *'noir* canon' commonly includes, for example, road movies (such as *Detour, Gun Crazy,* and *They Live By Night*), and small-town dramas (such as *Shadow Of A Doubt, The Reckless Moment, Beyond The Forest*). A significant number of *noir* films are also located outside the US – such as *Calcutta, Macao,* the Caribbean-set *The Bribe* and several films set in Latin America (*Gilda, Ride The Pink Horse, Cornered* and *Touch Of Evil*). *Film noir* also features a variety of period dramas, including *The Suspect* (set in Victorian London), *The Tall Target* (a thriller about an attempt to assassinate Abraham Lincoln), *Reign Of Terror* (a thriller set during the French Revolution), and *Pursued* (a *noir* Western). Nevertheless, Hollywood *film noir* remains most firmly associated with its distinctive imagining of the cultural and psychic topographies of the mid-century American big-city. From the 1970s, Hollywood cinema has extensively 'recycled' *noir* motifs and stylistics, and although a number of these neo-*noir* thrillers have made the original *noir* critique of the city more explicit (as is the case with *The Long Goodbye, Chinatown* and *Blade Runner*) many other films have shifted the focus from the city to dark hinterlands of America (in films as diverse as *The Hot Spot, After Dark, My Sweet, Blue Velvet, Red Rock West* and *One False Move*).

2   *Black Mask* published the first hard-boiled detective stories, by Carroll John Daly and Dashiell Hammett, in 1923. Daly's hero, Race Williams, has been described as 'the prototype for the modern private eye' (Ruehlmann, 1984: 58).

3   For useful overviews of anti-urban discourses in US politics, philosophy and literature, see the essays by White and White (1973), Rourke (1973), Gillette (1987), and the first chapter of Beauregard (1993: 3–33).

4   As critic Robert Warshow described the urban world of Hollywood's 1930s gangster films, the *noir* city is a dark city of the cultural imagination (Warshow, 1977: 128). Paul Arthur notes that in the *noir* thrillers, the city is 'the sum of a fairly limited but resonant set of constantly invoked motifs. . . . Cities resemble one another more than they present imagistic evidence of their differences. The same types of public and commercial buildings, vehicular structures, and open areas are employed in the schematic rendering of diverse urban venues. Among other salient qualities, there is a consistent stress, even for relatively "low" cities such as Los Angeles, on the vertical arrangement of space – locations are defined as above or below' (Arthur, 1990: 19, 222–3).

5   *It Happened One Night, Mr Deeds Goes To Town, You Can't Take It With You, Mr Smith Goes to Washington,* and *Meet John Doe* – on most of which Capra collaborated with New York playwright Robert Riskin.

6   To make Pottersville an instantly legible reverse-image of Bedford Falls, Capra's film accesses not only the city imagery of the *noir* thrillers but also the traumatic dislocation of (male) identity that is one of their predominant narrative obsessions. As I have considered elsewhere

(Krutnik, 1991: 86–91), the tough *noir* thrillers that descended from hard-boiled crime-fiction developed a distinctive form of male melodrama.

7   For a stimulating exploration of the role of 'homelessness' in *film noir*, see MacCannell (1993).

8   Based on a story by Philip Van Doren Stern that was initially published as a Christmas card pamphlet (Maland, 1995: 132–3; McBride, 1992: 510), *It's A Wonderful Life* was also released to coincide with the 1946 Christmas season. Robin Wood notes that, despite the upbeat sentiments of the film's ending, 'what is finally striking about the film's affirmation is the extreme precariousness of its basis; and Potter survives without remorse, his crime unexposed and unpunished' (Wood, 1986: 66). (The Bailey Building & Loan plunges into disaster after Potter opportunistically steals $8,000 from George's childlike, absent-minded Uncle Billy (Thomas Mitchell).

9   As Kasinitz stresses (p. 11), the *Gemeinschaft* and *Gesellschaft* distinction nostalgically valorizes the folk community of the premodern era, with the concept of the *Gesellschaft* implicitly demonizing the bureaucratized mass societies associated with modern cities. However problematic the opposition between *Gemeinschaft* and *Gesellschaft* may be in practice, it clearly has substantial symbolic purchase – as is illustrated by Capra's film.

10   McBride underscores the political evasions of the film by contrasting it with a more socially-critical version scripted earlier by leftist writer Dalton Trumbo (McBride, 1992: 22).

11   Although it was a disappointing box-office performer on its first release, the film's nostalgic celebration of community helped it to achieve almost mythic significance as a family-film for the Reaganite 1980s (McBride, 1992: 522).

12   Two major female stars turned down Capra's offer to play Mary Bailey: Ginger Rogers felt that 'the woman's role was such a bland character', and Jean Arthur said she 'wouldn't have liked to have been that girl, I don't think she had anything to do. It was colourless. You didn't have a chance to be anything' (quoted in McBride, 1992: 525). The more rigorously contained orbit of the women of this Capra film may be driven by contemporary anxieties concerning gender roles, spurred by the postwar readjustment of domestic priorities and the ejecting of women from the employment they had been encouraged to take during the war. Donna Reed's playing of Mary Bailey foreshadows the iconic American housewife of her eponymous TV sitcom of the late 1950s and early 1960s. Mary is doubled, somewhat nervously, with Violet (Gloria Grahame), a flirtatious good–bad girl who represents for George the (repressed) possibility of a more vicariously erotic relationship; in Pottersville, without his paternalistic protection, she ends up a prostitute. These two women serve as important focuses for George's dilemma: Violet suggests a teasing possibility of release (which he must continually deny himself) while Mary incarnates his familial, hence communal, duty.

13   Where the male protagonists of *Mr Deeds* and *Mr Smith* are relatively unified figures who are fighting against external forces, George Bailey's conflict is both

external (the struggle against the selfish Potter) and internal (the struggle against the Potter within) (Maland, 1995: 140). Potter himself makes the clear the connection between George and himself: 'You once called me a warped, frustrated old man – now you're a warped, frustrated young man.' Kaja Silverman provides a detailed and nuanced analysis of George's struggle against his frustrating legacy of cultural indebtedness – which, she argues, requires accepting and enjoying his enforced 'feminization' (Silverman, 1981: 17–18).

14   Some sampled moments: *Dead Reckoning* opens with the bruised and battered Rip Murdock (Humphrey Bogart) fleeing from his assailants through the streets, finding temporary refuge in a church; *Possessed* begins with the psychically-tormented Louise (Joan Crawford) roaming senselessly through the streets in search of the lover who has betrayed her; in *The Reckless Moment*, bourgeois housewife Lucia Harper (Joan Bennett) voyages into a squalid, poverty-wracked city to pawn jewellery to buy off the blackmailers who threaten her family; in *Out of The Past* private-eye Jeff Markham (Robert Mitchum) scrambles through the streets of San Francisco in a futile attempt to thwart a murder frame-up; in *Night and The City* small-time hood Harry Fabian (Richard Widmark) scurries about London in a failed bid to shake off the gangsters who are determined to kill him.

15   The crisis of self-figuration is rendered most literally in *noir* amnesiac narratives, such as *Street of Chance*, *Somewhere in The Night*, and *Crack-Up* (Krutnik, 1991: 132–3).

16   At the same time as the classic Western story enacts a cultural drama in which lawless forces are vanquished to make white American civilization possible, it stages a drama of male individualism that sits uncomfortably with the ostensible validation of community and manifest destiny. John Ford's *My Darling Clementine*, which appeared in the same year as *It's A Wonderful Life*, offers a self-consciously mythic treatment of these issues. Like Capra's film, Ford's Western presents a stylized, symbolic reassertion of the small-town community as the crucible of American social values, and it similarly uses a vision of the dark urban world as the antithesis of the community ideal – Tombstone is first shown as a night-time world of excess and chiaroscuro lighting (Wood, 1986: 63). The transition from lawlessness to order depends upon the intervention of a male hero, Wyatt Earp (Henry Fonda), who confronts and expels the forces of barbarism – principally incarnated in the all-male Clanton family (the demonic double of the Earp brothers). Although Earp offers himself as the champion of the community, he is not himself bound by it – and at the end of the film, he rides off back into the wilderness whence he came. Where Ford's Western validates the mythic promise of the American community by freezing history at that moment just before the West is tamed, when there was still a wilderness to escape into, *noir* fictions offer a more jaundiced view of the city as the apex of America's progress.

17   Chandler's first novel, *The Big Sleep* brings out these problems clearly: in Marlowe's hostility towards the gay relationship between Geiger and Carol Lundgren; in his own idealization of

romantic Irishman Rusty Regan; and in the savage torrent of sexual disgust inspired by the nymphomaniacal Carmen Sternwood in Chapter 24 (Chandler, 1948: 150–5). Howard Hawks' film adaptation, released in 1946, transforms Marlowe substantially by making him a man who is far more at ease with his sexuality.

18   For an indepth consideration of the relation between narration and subjectivity in *film noir*, see Telotte (1989).

19   Warner Brothers' 1941 adaptation of Hammett's *The Maltese Falcon* illustrates this well: with the exception of a brief establishing view of the San Francisco cityscape at the start, the film largely consists of a succession of indoor scenes in which the detective, Sam Spade (Humphrey Bogart) spars with his mysterious antagonists; the brief outdoor sequences are also filmed on studio sets. The restricted range of settings creates a pervasive interiority that reinforces visually the fact that Spade's allegiances have narrowed down to the self. Writing of Dashiell Hammett's source novel, William Marling says that 'Most of the action takes place in the confined spaces of apartments. . . . But that turns out to be one reason why it made such good *film noir*: interiors. Lack of a world elsewhere, of sunshine, can be emphasized. There are few guns or cars in the novel, but many telephone calls, newspapers and doorways, though none of these egress on a world elsewhere' (Marling, 1993: 81).

20   In the immediate postwar period, inflating costs of production encouraged a move out of the fixed-site studio. Twentieth-Century Fox made a series of crime films shot on location – including

The House On 92nd Street and *Call Northside 777* – which exploited the spectacle of the 'real city' (Lafferty, 1983: 22–6). Many more thrillers of the period combined location-footage with studio-shot material.

21   *When Strangers Marry* was made by the King Brothers, an independent production team signed to Monogram, one of the larger 'poverty row' companies specializing in B-product. Shot in ten days for under $50,000 (Nash and Ross, 1987: 3797), it was one of the films described by company president Steve Broidy as 'nervous A's', i.e. high-quality low-budget features designed to attract good critical notices and then hopefully to secure the kind of box-office percentage deals granted to A-films, rather than the flat rental fees usually reserved for B-product (McCarthy and Flynn, 1975: 271). Paul Kerr suggests that during the wartime cinema boom, even B-films were subject occasionally 'to press reviews and trade shows. Relatively rapidly B films were encouraged to become increasingly competitive, compulsorily different, distinctive. What had previously, perhaps, been a rather static aesthetic and occupational hierarchy between Bs and As became suddenly more flexible' (Kerr, 1983: 50). Unable to book a prestige Manhattan theatre for the opening of *When Strangers Marry*, Monogram did not succeed in obtaining a percentage-deal until the next King Brothers feature, *Dillinger* (Miller, 1987: 310–11). Nevertheless, the film did very good business and received excellent reviews, as the following excerpts suggest: 'a taut psychological thriller . . . Psych mood is cleverly sustained throughout for good atmosphere' (*Variety*, 1944: 13); 'taut, artistically handled mystery melodrama

... [which] works cleverly in an atmosphere laden with suspense. ... The stars' performance, direction and photography are much above average' (*Kine Weekly*, 1945: 20); 'I have seldom, for years now, seen one hour so energetically and sensibly used in a film. Bits of it, indeed, gave me a heart-lifted sense of delight in real performance and perception and ambition which I have rarely known in any film context since my own mind, and that of moving-picture making, were both sufficiently young' (Agee, 1958: 155). Championed at the time by respected cultural voices Manny Farber, Orson Welles and James Agee, the film has subsequently received remarkably little attention in the voluminous writings on *film noir* – although Don Miller refers to it as 'unquestionably the finest B film made' (Miller, 1987: 310).

22 '[D]uring the first two thirds of the twentieth century, New York, more than any other location on the planet, epitomized the forces of modernity. The very names of its streets – Madison Avenue, Wall Street and Broadway – became synonymous with the structures of mass communications and persuasion, finance, capitalism and popular culture that have defined the modern way of life. New York dominates the conceptual and symbolic vocabulary with which we think of the modern city' (Kasinitz, 1995: 86).

23 Their scheme comes unstuck, however, because they do not have equal investments in the transgression. As a male, Walter is more secure in the patriarchal system he revolts against – and as an insurance salesman he knows that taking risks is itself part of the system. But Phyllis plays a more dangerous game –

and, because she strikes against the system from which she is excluded, her scenario of revolt has to be more rigidly contained, and disarticulated, within the text.

24 The *femme fatale* is alluring for the hero because she both inspires and embodies a fantasy of revolt against containment. She must ultimately pay the price – not just for her own transgression but for his, too – through a ritualistic annihilation. In *Double Indemnity*, *Out of The Past*, *Dead Reckoning*, *The Lady From Shanghai*, *The Strange Love Of Martha Ivers* and *Gun Crazy*, the perverseness of the *noir* woman is affirmed when she wields a gun – unquestionable proof of her phallic desire, which simultaneously renders her amenable to containment and eradication.

25 Although, as Elizabeth Cowie (1993: 132–7) has cautioned, there are many notable hybrids of the masculine *noir* crime thriller and the women's picture melodrama, including *Mildred Pierce*, *The Reckless Moment*, *Possessed* and *Nora Prentiss*.

26 Similar scenes, in which a character is plagued by projected faces and/or voices, feature in *Scarlet Street*, *The Gangster* and *Bewitched*. Such cases illustrate how in *film noir* the materiality of image and sound can give way to an expressionist mode of representation that transforms the urban world into a projected landscape of nightmare and shadows.

27 These 1940s films have received much attention from feminist critics – cf. Doane, (1987: 123–75; Waldman, 1984: 44–7); the account of *The Secret Beyond The Door* in Cowie (1993: 145–59).

28   For example, the spectacle of a lone woman terrorized by the mysterious dangers of the city is found in numerous films, including *The Stranger On The Third Floor*, *Cat People*, *The Leopard Man*, *Phantom Lady*, *Possessed*, *The Reckless Moment* and *Bewitched*.

29   *The Monthly Film Bulletin* (1945: 23) also complained about the film's 'somewhat illogical' story development.

30   At Coney Island, Paul and Millie visit the sideshow of Hugo, the Mental Marvel, who boasts that he can discern anyone's occupation from their appearance. He places Millie straight away as a newlywed, but he fails to identify Paul correctly – guessing erroneously that he is a reporter, a lawyer, a bank worker. Such tensions between surface and depth, familiarity and strangeness, the manifest and the latent, are central to *film noir*.

31   Sunk in abjection, Paul lacks direction and motivation. Millie repeatedly takes the dominant role in his escape attempt and it is she, too, who – like the women in *The Stranger on the Third Floor* and *Phantom Lady* – eventually proves the innocence of the man she loves.

32   After 1890, the migration of African–Americans from the urban South accelerated (Chudacoff, 1975: 90). By the 1940s, they were a prominent minority in the industrial cities of the North and West, but the black figures most common in 1940s *film noir* are the maids of white women (as in *Mildred Pierce*, *Dead Reckoning*, *The Lady From Shanghai*). However, *Out Of The Past* features an unusual scene in which the detective visits an all-black club to question the maid of the gangster's moll he is searching for. Although its main point is to emphasize the ease with which the private-eye can move through the city's various worlds, this scene is nonetheless free of Hollywood's familiar racial condescension. Despite the limited presence of African–American characters, black music is frequently appropriated as a signifier of libidinal excess – the frantic jazz sequences of films such as *Among The Living*, *Phantom Lady*, and *D.O.A.* forge a familiar connection between the 'dark' forces of the unconscious and the otherness of race. Deborah Thomas suggests the broader difficulty that *film noir* has with ethnicity: 'the characteristic anxiety provoked by the contemporary urban setting of *film noir* has its roots, at least in part, in a response to the waves of immigration of the late nineteenth and early twentieth centuries which seemed to make the city no longer the locus of American 'civilization' (a native version of white Anglo-Saxon Protestantism) but rather of an antithetical "otherness"' (Thomas, 1992: 61). *The Maltese Falcon* is an especially clear example of the anxiety she speaks of: as William Luhr (1990: 17–19) has considered, Sam Spade's adventure moves him through a San Francisco overrun with foreign invaders. Confronting the motley collection of outsiders, sexual 'deviants' and a *femme fatale*, Spade battles to assert the superiority of his own brand of isolationist American manhood – his victory secured when he exposes the exotic jewelled bird of the title as lacking in real value.

33   Paul Schrader (1986: 170) argues that *film noir* 'is not defined, as are the western and the gangster genres, by conventions of setting and conflict, but rather by the more subtle qualities of tone and mood'. However, conventions of setting and conflict are not as marginal as he maintains – as the above reading of *When Strangers Marry* makes plain, *film noir*'s tone or mood derives not just from the resonant scenes themselves but also from the manner in which the narrative juxtaposes them.

FILMS CITED

After Dark, My Sweet (d. James Foley, 1990)

Among The Living (d. Stuart Heisler, 1941)

Bewitched (d. Arch Oboler, 1945)

Beyond The Forest (d. King Vidor, 1949)

The Big Heat (d. Fritz Lang, 1953)

The Big Sleep (d. Howard Hawks, 1946)

Blade Runner (d. Ridley Scott, 1982)

Blue Velvet (d. David Lynch, 1986)

The Bribe (d. Robert Z. Leonard, 1949)

Calcutta (d. John Farrow, 1947)

Call Northside 777 (d. Henry Hathaway, 1948)

Cat People (d. Jacques Tourneur, 1942)

Chinatown (d. Roman Polanski, 1974)

Cornered (d. Edward Dmytryk, 1945)

Crack Up (d. Irving Reis, 1947)

D.O.A. (d. Rudolph Mate, 1949)

The Dark Corner (d. Henry Hathaway, 1946)

Dark Passage (d. Delmer Daves, 1947)

Dead Reckoning (d. John Cromwell, 1947)

Detour (d. Edgar G. Ulmer, 1945)

Dillinger (d. Max Nosseck, 1945)

Double Indemnity (d. Billy Wilder, 1944)

Experiment Perilous (d. Jacques Tourneur, 1944)

The Gangster (d. Gordon Wiles, 1947)

Gaslight (d. George Cukor, 1944)

Gilda (d. Charles Vidor, 1946)

Gun Crazy (d. Joseph H. Lewis, 1950)

The Hot Spot (d. Dennis Hopper, 1990)

The House On 92nd Street (d. Henry Hathaway, 1945)

The House on Telegraph Hill (d. Robert Wise, 1951)

It Happened One Night (d. Frank Capra, 1934)

It's A Wonderful Life (d. Frank Capra, 1946)

The Killers (d. Robert Siodmak, 1946)

Kiss Of Death (d. Henry Hathaway, 1947)

The Lady From Shanghai (d. Orson Welles, 1948)

Laura (d. Otto Preminger, 1944)

The Leopard Man (d. Jacques Tourneur, 1943)

The Long Goodbye (d. Robert Altman, 1973)

Macao (d. Joseph von Sternberg, 1952)

The Maltese Falcon (d. John Huston, 1941)

Meet John Doe (d. Frank Capra, 1941)

Mildred Pierce (d. Michael Curtiz, 1945)

Mr Deeds Goes To Town (d. Frank Capra, 1936)

Mr Smith Goes To Washington (d. Frank Capra, 1939)

My Darling Clementine (d. John Ford, 1946)

My Name Is Julia Ross (d. Joseph H. Lewis, 1945)

Night And The City (d. Jules Dassin, 1950)

Nora Prentiss (d. Vincent Sherman, 1947)

One False Move (d. Carl Franklin, 1992)

Out Of The Past (d. Jacques Tourneur, 1947)

Phantom Lady (d. Robert Siodmak, 1944)

Pickup On South Street (d. Sam Fuller, 1953)

Possessed (d. Curtis Bernhardt, 1947)

The Postman Always Rings Twice (d. Tay Garnett, 1946)

Pursued (d. Raoul Walsh, 1947)

Rebecca (d. Alfred Hitchcock, 1940)

The Reckless Moment (d. Max Ophuls, 1949)

Red Rock West (d. John Dahl, 1992)

Reign Of Terror (d. Anthony Mann, 1949)

Ride The Pink Horse (d. Robert Montgomery, 1947)

Rope (d. Alfred Hitchcock, 1948)

Scarlet Street (d. Fritz Lang, 1945)

The Secret Beyond The Door (d. Fritz Lang, 1947)

Shadow Of A Doubt (d. Alfred Hitchcock, 1943)

Shockproof (d. Douglas Sirk, 1948)

Somewhere In The Night (d. Joseph Mankiewicz, 1946)

The Strange Love Of Martha Ivers (d. Lewis Milestone, 1946)

The Stranger on the Third Floor (d. Boris Ingster, 1940)

Strangers On A Train (d. Alfred Hitchcock, 1951)

Street of Chance (d. Jack Hively, 1942)

The Suspect (d. Robert Siodmak, 1944)

The Tall Target (d. Anthony Mann, 1951)

They Live By Night (d. Nicholas Ray, 1949)

They Won't Believe Me (d. Irvin Pichel, 1947)

Touch Of Evil (d. Orson Welles, 1958)

When Strangers Marry/Betrayed (d. William Castle, 1944)

You Can't Take It With You (d. Frank Capra, 1938)

You Only Live Once (d. Fritz Lang, 1937)

REFERENCES

Agee, J. (1958) '*When Strangers Marry*', in *The Nation* (7 April 1945) (review), in *Agee On Film*, McDowell-Obolensky, New York.

Annenberg (1994) '*Film Noir*', documentary in the video series *The American Cinema*, from the Annenberg CPB Project.

Arthur, P. (1990) *Shadows on the Mirror*: Film Noir *and Cold War America* (PhD thesis, New York University, 1985), University Microfilms International, Ann Arbor, Michigan.

Beauregard, R. A. (1993) *Voices of Decline: The Postwar Fate of US Cities*, Blackwell, Cambridge, Mass.

Chandler, R. (1948) *The Big Sleep* [1939], Penguin Books, Harmondsworth.

Chandler, R. (1973a) 'The Simple Art Of Murder' [1944], in *Pearls Are A Nuisance*, Penguin Books, Harmondsworth, 181–99.

Chandler, R. (1973b) 'Introduction' [1950], in *Pearls Are A Nuisance*, Penguin Books, Harmondsworth, 7–10.

Chudacoff, H. P. (1975) *The Evolution of American Urban Society*, Prentice-Hall, Englewood Cliffs, NJ.

Cowie, E. (1993) '*Film Noir* and Women', in J. Copjec, (ed.) *Shades of Noir: A Reader*, Verso, London, 121–65.

Crowther, B. (1953) '*Pickup On South Street*' (review), *New York Times*, 18 June, p. 38.

Doane, M. A. (1987) *The Desire To Desire: The Woman's Film of the 1940s*, Macmillan, London.

Dyer, R. (1977) 'Homosexuality and *Film Noir*', *Jump Cut* 16 November, 18–21.

Framework, Editorial Board (1976) 'The Family in *The Reckless Moment*', *Framework* 4 Autumn, 21–4.

Gillette, H. Jr. (1987) 'The City in American Culture', in H. Gillette and Z. L. Miller (eds) *American Urbanism: A Historiographical Review*, Greenwood Press, Westport, Conn. 27–48.

Harvey, S. (1978) 'Woman's Place: The Absent Family of *Film Noir*', in E. Ann Kaplan (ed.) *Women in Film Noir*, British Film Institute, London, 22–34.

Higham, C. and Greenberg, J. (1968) *Hollywood in the Forties*, Tantivy Press, London.

Houseman, J. (1947) 'Today's Hero: A Review', *Hollywood Quarterly* 2 (2), 161–3.

Kasinitz, P. (1995) *Metropolis: Centre and Symbol of Our Times*, Macmillan, London.

Kerr, P. (1978) 'Out of What Past? The "B" *Film Noir*', *Screen Education* 32–3, 45–65.

Kerr, P. (1983) 'My Name is Joseph H. Lewis', *Screen* 24 (4), 48–66.

Kinematograph Weekly (1945) '*When Strangers Marry*' (review), *Kinematograph Weekly*, 25 January, p. 20.

Knight, S. (1988) '"A Hard Cheerfulness": An Introduction to Raymond Chandler', in B. Docherty, (ed.) *American Crime Fiction: Studies in the Genre*, St Martin's Press, New York 71–87.

Krutnik, F. (1991) *In a Lonely Street*: Film Noir, *Genre, Masculinity*, Routledge, London.

Lafferty, W. (1983) 'A Reappraisal of the Semi-Documentary in Hollywood', *The Velvet Light Trap* 20 (Summer), 22–9.

Lipsitz, G. (1990) *Time Passages: Collective Memory and American Popular Culture*, University of Minnesota Press, Minneapolis.

Luhr, W. (1990) 'Tracking *The Maltese Falcon*: Classical Hollywood Narration and *Film Noir*', in P. Lehman (ed.) *Close Viewings*, University of Florida Press, Tallahassee, 7–22.

McBride, J. (1992) *Frank Capra: The Catastrophe of Success*, Faber and Faber, London.

MacCannell, D. (1993) 'Democracy's Turn: On Homeless *Noir*', in J. Copjec (ed.) *Shades of Noir: A Reader*, Verso, London, 279–97.

McCarthy, T. and Flynn, C. (eds) (1975) *Kings of the B's: Working Within the Hollywood System*, E. P. Dutton, New York.

Maland, C. (1995) *Frank Capra*, Twayne Publishers, New York.

Marling, W. (1993) 'American *Roman Noir* and *Film Noir*', *Literature/Film Quarterly*, 21 (3), 178–93.

Monthly Film Bulletin (1945) '*When Strangers Marry*' (review), *The Monthly Film Bulletin* 12 (134), 23.

Miller, D. (1987) *B Movies*, Ballantine Books, New York.

Muller, T. (1993) *Immigrants and the American City*, New York University Press, New York.

Nash, J. R. and Ross, S. R. (1987) '*When Strangers Marry*', *The Motion Picture Guide, Volume 9*, Cinebooks, Chicago.

Rourke, F. E. (1973) 'Urbanism and American Democracy', in A. B. Callow, (ed.) *American Urban History: An Interpretive Reader, With Commentaries*, Oxford University Press, New York, 426–40.

Ruehlmann, W. (1984) *Saint With a Gun: The Unlawful American Private Eye*, New York University Press, New York.

Schrader, P. (1986) 'Notes on *Film Noir*' [1972], in B. K. Grant, (ed.) *Film Genre Reader*, University of Texas Press, Austin, 169–82.

Silverman, K. (1981) 'Male Subjectivity and the Celestial Suture: *It's a Wonderful Life*', *Framework* 14 (Spring), 16–22.

Telotte, J. P. (1989) *Voices in the Dark: The Narrative Patterns of Film Noir*, University of Illinois Press, Urbana.

Thomas, D. (1992) 'How Hollywood Deals With The Deviant Male', in I. Cameron (ed.) *The Movie Book of Film Noir*, Studio Vista, London, 59–70.

Variety (1944) '*When Strangers Marry*' (review), *Variety*, 29 November, p. 13.

Waldman, D. (1984) '"At Last I Can Tell It To Someone": Feminine Point of View and Subjectivity in the Gothic Romance Films of the 1940s', *Cinema Journal* 23 (2), 44–50.

Warshow, R. (1977) 'The Gangster As Tragic Hero' [1948], in W. M. Hammel (ed.) *The Popular Arts in America: A Reader*, 2nd edition, Harcourt Brace Jovanovich, New York, 125–30.

White, M. and White, L. (1973) 'The American Intellectual versus the American City', in A. B. Callow (ed.) *American Urban History: An Interpretive Reader, With Commentaries*, Oxford University Press, New York, 416–26.

Wood, R. (1986) 'Ideology, Genre, *Auteur*' [1977], in B. K. Grant, (ed.) *Film Genre Reader*, University of Texas Press, Austin, 59–73.

5

URBAN CONFIDENTIAL
the lurid city of the 1950s
•

Will Straw

INTRODUCTION

The first few minutes of *New Orleans After Dark* (1957) offer one of the most glaringly inept opening sequences in the American cinema of the 1950s. As the film begins, prior to the appearance of its credits, we are shown a woman in a nightclub, singing at a piano. As she sings – a song whose lyrics inventory the range of human types to be found along New Orleans' Bourbon Street – we see a montage of images plucked from the narrative which follows, most of them involving violence against female burlesque dancers. Before the song's conclusion, the film shifts to an oddly flat, stagy interior tableau in which a policeman bids farewell to his wife and son before heading off to the night shift. A voice-over notes that "[t]his film could be about dedicated police officers anywhere." Brief, on-location images of downtown New Orleans follow, themselves succeeded by scenes of brief banter between patrol-men and a streetwalker; already our sense of chronology has become confused. Over more scenes of the street, the words of the film's title and credits are superimposed, in lettering which recalls the typescripts used in scandal magazines of the time. As the credits conclude, we move into a burlesque club, and into what we may assume is the beginning of the film's narrative.

The generic threads interwoven here suggest, simultaneously, the traditions of the semi-illicit stag film, the police procedural, the semi-documentary instruction film, the vice exposé movie and the low-budget mystery. Each of these had assumed a relative solidity, in the cinema of the 1950s, as recognizably distinct means of representing the American city. In the confusion which results from their mixing, they identify *New Orleans After Dark* as the symptom of a transition – one in which the distinctiveness of these generic traditions, and of the industrial structures which sustained them, was beginning to dissolve. What remains is the sense of *New Orleans After Dark* as an obscure, marginal film, a sign of the decaying of several threads within the postwar American cinema in which the city had been prominent.

New Orleans After Dark is one of the last of a cycle of American films which began in the early 1950s and had concluded by the end of the decade. The titles

5.1   *Las Vegas Shakedown* (Allied Artists, 1955)
Courtesy of BFI Stills, Posters and Designs.

of these films named cities and promised the revelation of secrets about them. Others in this cycle include *The Phenix City Story* (1955), *Portland Expose* (1957), *New Orleans Uncensored* (1955), *Kansas City Confidential* (1952), *New York Confidential* (1955), *Las Vegas Shakedown* (1955; Plate 5.1), *The Houston Story* (1956) *Chicago Syndicate* (1955), *Miami Expose* (1956) and *Inside Detroit* (1955). Most of these films claimed, in their opening sequences or in the posters which advertised them (Plate 5.2), some link to a real-life investigation of municipal vice and corruption.[1] Their narratives, nevertheless, are secondary to their cataloguing of vice, and to the formal organization of these films as sequences of scenes in night-clubs, gambling dens and along neon-lit streets. (Indeed, as the cycle winds down, by 1958 or 1959, the frequency of journalistic tie-ins declines.)

This cycle of city exposé films was inspired, in its initial phases, by the Senate hearings on municipal corruption chaired by Senator Estes Kefauver of Tennessee. By the time of *New Orleans After Dark*, the influence of that political event persists only in the practice of incorporating city names in a film's title, and in a brief reference to a grand jury investigation of a narcotics racket. However, that film is inseparable

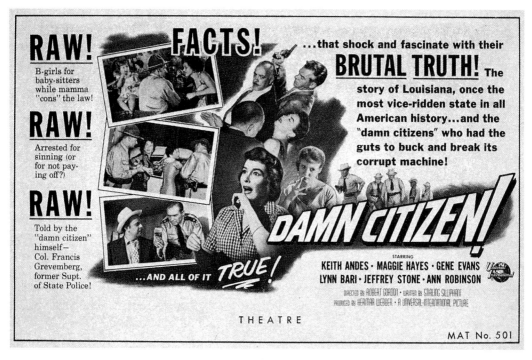

5.2   One of the posters advertising *Damn Citizen*

from the backdrop of a culture of urban exploitation and exposé which flourished in the middle years of the decade. As a nocturnal, urban crime film, *New Orleans After Dark* is one moment in the unravelling of *film noir*, but this is *noir* after the major studios, prestige stars and canonical directors have left. Like many of the films in this cycle, *New Orleans After Dark* was shot on location, and some of its principal characters were played by non-professional locals, including members of the New Orleans police force. (This film's poster claims that it was "[f]ilmed in the sin spots where it happened!") These elements situate *New Orleans After Dark* within a lineage which reaches back to *The House on 92nd Street* (1945) or *13 Rue Madeleine* (1946), but the status of these semi-documentary elements has by now shifted. Originally the mark of a moral seriousness, they now stand for the obscurity of production conditions, for the move towards regionalist, exploitational film making practices which will continue throughout the late 1950s and early 1960s.

"THE REST ARE ALL CLEVELAND"

The popular culture of the US has, of course, been marked by recurrent cycles of urban exposé – of literary, journalistic or other works in which one finds a variable balance between the ameliorative impulse towards documentation and the exploitational imperative to produce moments of textualized sensation.[2] The cycle which concerns us here turns on the term "confidential," a term which moves to the centre of popular cultural discourse through a series of exposé books published (and sold in large numbers) in the decade following the Second World War. As it circulates, the term will be one (albeit disputed) influence on the hearings of a judicial committee of investigation, will provide the title for the most successful new magazine of the 1950s, and will nourish a series of films which mark the passage from postwar *film noir* to new cycles of cinematic exploitation.

What marks this intertextual space, in part, is the proliferation of urban sites which films, magazine articles and exposé books will posit as centers of vice and corruption to be exposed. New York, Chicago and other prominent cities will figure in many of these texts, but as the cycle unfolds there is a dispersion of attention outwards, towards medium-sized cities, regional capitals, and, in a variety of films, fictionalized versions of the mythically corruption-ridden "wide open" town. (See, for one such fictional location, *The Long Wait*, 1954.) This logic of dispersion mirrors the dissemination of public concern over municipal corruption throughout the 1950s, concern fueled by investigative journalism operating in innumerable local contexts. It will be shaped, as well, by the common patterns by which exploitational texts are given distinctiveness within a commercial and cultural field. Once particular generic frameworks have been established – the exposé film whose title names a city, a magazine column (in the periodical *Focus*) entitled "Your Town Confidential" – examples will proliferate in order that the production of differentiated texts may continue.

THE CITY CONFIDENTIAL BOOKS

In 1948, journalists Jack Lait and Lee Mortimer published *New York Confidential*, the first in a series of four co-authored volumes which offered the "low-down" on vice and corruption in contemporary US cities.[3] Lait brought to these books an association with the mythical traditions of Chicago journalism, in which he had begun his career, and personal fame for having been the first journalist to report on the murder of gangster John Dillinger in 1934. He had written several popular books on New York in the 1920s and 1930s, including *Broadway Melody* (1928) and, since 1936, had been editor of the *New York Mirror*. Mortimer, who had worked for Lait as a columnist since the 1930s, would become best known, in the late 1940s, as the journalist whom Frank Sinatra punched outside a nightclub.[4] The *Confidential* books were significant best-sellers, each running through several reprintings in paperback until a series of lawsuits in the mid-1950s served to discredit them (Plate 5.3).[5]

In the unfolding of this series, one finds evidence of the shift in postwar journalistic treatments of the US city to which Neal Gabler has alluded in his biography of Walter Winchell.[6] From first to last, the Lait and Mortimer books signal the

5.3   Cover of the paperback release of *Washington Confidential* (Dell Books, 1951)
Courtesy of Bantam, Doubleday, Dell Publishing Group Inc.

transition from a Damon Runyonesque celebration of the city as playground for lovable eccentrics to a more thuggish journalism preoccupied with exposing vice, corruption and moral decay. *New York Confidential* is faithful to a long tradition of writing which celebrates the encyclopaedic inexhaustibility of the monumental American city, offering it to the curious outsider's eye as merely the biggest of villages.[7] In the postwar period, this tradition would fragment, such that many of its elements would be reconfigured within a broader vision which linked municipal criminality to a generalized weakening of national moral fibre and, by implication (or direct accusation), to ill-preparedness against the communist threat. The sense of the city as folkloric, as distinctive because of its linguistic vernacular and endearing rituals, would give way to a sense of all cities as barometers of a generalized social rot.

Indeed, by the time of *U.S.A. Confidential*, the distinctiveness of American cities is seen, by Lait and Mortimer, to be withering, dissolved within a generalized mediocrity and corruption:

> We learned [in researching the book] that the conditions of vice, crime, graft and organized racketeering have spread over virtually the entire nation, rural as well as metropolitan, much of it even more revolting and alarming than the shameful corruption of the three major cities. The tax cheating which has come to general light is only the surface scum.

> It is a sin against the new world of mediocrity to be distinct or distinguished. We are in the chain-store, neon-lighted era. Almost every city looks the same. The same people all dress the same – kids as Hopalong Cassidy, men with loud sportshirts and Truman suits, women in slacks. Sometimes you can tell whether a trousered individual is a man or a woman only by the width of the buttocks. Only a few cities have individuality. They are the seaports, New York, New Orleans and San Francisco. Boston reeks of decay, and is not genteel. The rest are all Cleveland.
>
> (Lait and Mortimer 1952: x, 6)

The passage of these impulses into the American cinema is partial and relatively indirect. It is limited, at one level, by the persistence of a liberal reformism as the principal prism through which urban ills would be understood in the American cinema of the 1950s. The link between the *Confidential* books and urban exposé films of the time was mediated, as well, by the influence of the so-called Kefauver Committee hearings on organized crime and urban corruption. In their own account of the origin of these hearings, Lait and Mortimer claimed that Senator Estes Kefauver of Tennessee, looking for a book to put him to sleep, picked up a copy of *Chicago Confidential*, and was shocked into action by the revelations it contained.[8] The result was a joint committee of the Senate Judiciary and Commerce committees, which traveled the country in 1950 and 1951 looking for evidence of municipal corruption and the influence of organized crime on city governments.[9]

The more convincing account, offered by Moore, is that Lait, Mortimer and Kefauver were all responding to a broader postwar concern over municipal corruption, a concern evident in the postwar revival of citizen's crime commissions and in the emergence of new press syndicates devoted to investigative reporting on vice and organized crime (Moore 1974: 38–9).

The Kefauver Commission held hearings in fourteen major cities, and heard testimony from over 800 witnesses. In one case, a potential witness was murdered; in several others, threats to witnesses were received and publicized. Television coverage of the Kefauver Committee's hearings began in January of 1951, and, by the time the Committee reached New York, was attracting a morning television audience several times the norm (Moore 1974: 184–5). Indeed, historians of US media have claimed that the Kefauver hearings represented the first significant use of television for the broadcasting of judiciary or legislative proceedings (Derr 1986). More significantly, for our purposes, the notion that crime was something to be "uncovered" (rather than merely controlled, or analyzed for its social causes) would, in the aftermath of the hearings, shape a broad and varied corpus of popular cultural texts. The *Saturday Evening Post* published a four-part series, *What I Found In the Underworld*, which was ghostwritten for Kefauver. CBS broadcast a television series, *Crime Syndicate*, which used members of the Committee as narrators; the run of *Racket Squad*, on the same network, coincided with the period of the Committee's hearings.

The itinerary of the Kefauver Committee's deliberations would institute a structure with significant effects on a variety of popular cultural forms. Against those who advised him to remain in Washington and call witnesses to the Capitol, Kefauver decided to travel from city to city, calling local witnesses, cultivating relations with the local press, and purporting to deal with the histories of criminality and corruption specific to each locale. This helped to enshrine a conception of his project as a series of local investigations, each marked by its separate rhythms of dramatic intensity and given texture through the particularities of local speech patterns and power structures. As well, this fixation on local circumstances worked against the notion of municipal corruption as the product of a national conspiracy, a web-like crime syndicate – a notion which many, particularly Lait and Mortimer, thought should be the Commission's ultimate conclusion (Lait and Mortimer 1952: 22; Mortimer 1951: 521). The form of the Committee's report and of many of its spin-offs was that of a series of individual cases lacking a clear sense of coherent implications at the national level (e.g. US Congress, 1951). This would be mirrored in much of the literature and cinema of urban exposé produced in the wake of the hearings.

Of the urban crime films made while the Kefauver hearings were underway, *The Captive City* (1951) offers the most direct connection to the event (Plate 5.4). It tells the story of a journalist, working in a medium-sized city, who discovers pervasive corruption within municipal politics and flees for his life with the information. Its narrative culminates with the arrival of the central protagonist at the Committee's hearing rooms where he will tell his story. The film itself has, as

5.4 Lobby card for *The Captive City*
© United Artists, 1952.

an epilogue, a direct-address speech by Kefauver warning the viewer that gambling is a national problem. In terms of both its production values and its resistance to an exploitational logic, *The Captive City* may be seen as the most earnestly reformist and respectable of those films with direct links to the Kefauver Committee. *The Enforcer* (1951), whose narrative presumes the existence of a national murder-for-hire syndicate, is likewise linked to the Kefauver hearings, though its stylistic and narrative elements are more typical of other crime films of the period. *The Racket* (1951), while a remake of a 1928 film of the same name, acknowledges the existence of the Kefauver Committee through the intervention, in its concluding moments, of a State Crime Commission. None of the films in this first cycle incorporates the names of cities in their titles. Indeed, in their emphasis on the coming-to-consciousness of law enforcement officers or citizens, rather than the social roots of geographically specific corruption, they suggest that location is unimportant.

A number of other films produced in the early and mid-1950s refer directly to the Kefauver hearings or to the investigations undertaken in their wake. These

5.5   *Portland Expose* (Allied Artists, 1957)
Courtesy of BFI Stills, Posters and Designs.

include *The Phenix City Story*, *Portland Expose* (Plate 5.5), and *Kansas City Confidential*. Other films – *New York Confidential* and *Chicago Confidential* – while ostensibly based on the Lait and Mortimer books, clearly position themselves as coming after the Kefauver hearings, thereby benefitting from their publicity. Around 1955, a year in which the release of several of these films is concentrated, the incorporation of city names within titles has become a recognizably entrenched means by which producers claim a link to an emergent cultural formation centered on the exposure of urban vice and criminality. This tendency will be dispersed throughout the popular culture of the mid-1950s, manifest in additional films, such as *New Orleans Uncensored*, *New Orleans After Dark* and *Inside Detroit*, and in a variety of other texts and forms.

What is striking about this cycle of films is its rapid decline to levels of production and prestige which run counter to the monumental, nationally resonant quality of the hearings themselves. Those which come late in the cycle will bear ever more blatant signs of an exploitational logic. While almost all make reference to journalistic sources in their advertising, they simultaneously foreground an imagery of corruption and illicit sexuality, promising, as does the poster for *Portland Expose*, to be "[*n*]*akedly* shocking on the screen!" Arguably, their relationship to the Senate hearings is mediated by the intervening wave of exposé and scandal magazines, to which I will turn in a moment. More significantly, perhaps, the unfolding of this cycle follows the ongoing decentralization of the Hollywood studio system: many of these films are filmed on location, in mid-sized cities, and use unknowns or performers of low status. As suggested earlier, this is no longer evidence of the deliberate and reformist semi-documentary turn of the immediate postwar period. Rather, by the time of such films as *New Orleans After Dark*, it will stand for the resurgence of regional film making practices and marginal distribution and exhibition circuits. Most of these films are about peripheral geographical locations, and their own thematic and industrial obscurity works to block their participation in any generalized, moral panic over organized crime. (Indeed, their documentary-like, procedural sequences are typically quite brief, and the films are, for the most part, dominated by lurid interiors of gambling dens or nightclubs.)

Organizational similarities link the later Lait and Mortimer books and the report of the Kefauver Commission. There is a similar emphasis in both on the ease with which one can procure alcohol or the services of a prostitute in any urban locale, the same fixation on showing that patterns of corruption have been replicated from one city to another. Indeed, both the Commission's report and *U.S.A. Confidential* leave the impression that all cities are identical, at least after dark. The accumulation of detail about each case, it may be argued, serves principally to meet the needs of the publishing market or the political headline-grabber. This construction of a national space of relatively undifferentiated corruption will be mirrored in much of the paperback fiction of the 1950s (such as the novels of Jim Thompson), whose locales are often unidentified but characterized by patterns of corruption now assumed to be replicated throughout the United States. It will help organize magazine exposés of such corruption as minor local variations on broadly established themes.

ROGUE COPS

Arguably, the proceedings of the Kefauver Committee are one influence on one of the most significant developments within postwar crime fictions: the emergence of the urban policeman as a complex, often tragic figure. A variety of threads within the postwar American cinema enact a transformation of police characters from undeveloped ethnic figures of ridicule or inconsequence to fictional persona whose characterological density is the pivot around which narratives frequently turn. These threads include the emergence of the police procedural film (such as *The Naked City*, 1948 and *The Tattooed Stranger*, 1950), a series of *films noirs* which undertake the psychologization of policemen as bearers of class resentment or disgust at urban degradation (*Between Midnight and Dawn*, 1949; *On Dangerous Ground*, 1952; *Shield for Murder*, 1954), and gangster films within which police corruption is figured as a central concern (*The Big Heat*, 1953; *Rogue Cop*, 1954). In popular literature, as well, the novels of such writers as William P. McGivern would center on a series of corrupt policemen acting in complicity with racketeers (e.g. McGivern 1956).

The familiarity which such themes have acquired over the last forty years should not obscure the extent to which the notion of the policeman as a figure of dramatic complexity is a product of postwar fictions. The policeman will come to stand, in film cycles which continue through the present, as the locus of fictions which address the relationship between crime, social order and individual morality, and the tensions between professional, bureaucratic and political authority. These tensions between different orders of authority are what mark the difference of the police film from the private eye tradition, and their emergence as fictional themes in the 1950s is inseparable from the preoccupation with the institutions of law enforcement which the Kefauver hearings helped to nourish.

This transformation of the figure of the policeman is concurrent with a comparable complexification of the figure of the nightclub singer or prostitute. Indeed, in a variety of crime films of the 1950s, there emerges an image of a *demi-monde* in which the policeman and the woman of the night both assume greater density as fictional persona than was common in the films of earlier decades. Writing of the street woman of the nineteenth century, as seen through the eyes of the strolling male writer, but in terms applicable to the urban crime film of the 1950s, Deborah Epstein Nord notes that

> [t]he sexually tainted woman can stand variously as an emblem of social suffering or debasement, as a projection of or analogue to the male stroller's alienated self, as an instrument of pleasure and partner in urban sprees, as a rhetorical and symbolic means of isolating and quarantining urban ills in the midst of an otherwise buoyant metropolis, or as an agent of connection and contamination.
>
> (Nord 1991: 353–4)

In the corrupt police thrillers of the 1950s, the problematic of the sexually tainted woman turns less and less on the alternatives (central to the private eye film) of betrayal, redemption and the status of individual nobility in a world of seductive temptations. Rather, in a tradition that will extend into the 1970s and 1980s, her presence within fictions poses the question of whether any relationships are possible outside the homosocial bonds of the police force and criminal syndicates who, virtually alone, remain to inhabit the inner city. As the exposé culture of the 1950s develops, there is a marked shift from an earlier preoccupation with gambling to a fixation on the figure of the prostitute or B-girl. In cinema, this marks the decline of the gangster film revival which had begun, in tandem with the Kefauver hearings, in the early 1950s. The fixation on the sinful urban woman will continue through a more blatantly exploitational cycle of films which includes such titles as *The Female Jungle* (1956) and *The Violent Years* (1956).

THE "LOW-DOWN" MAGAZINES

The passage of the word "confidential," from the titles of the Lait and Mortimer books to that of what Tom Wolfe called the "most scandalous scandal magazine in the history of the world," is relayed through the hearings of the Kefauver Committee on municipal corruption (Wolfe, quoted in Gabler 1994: 468). Various accounts agree that Robert Harrison, who had built a career publishing pin-up magazines during and after the Second World War, saw the Kefauver hearings on television and realized, as he himself put it, that "finally the public had become educated to the fact that there was excitement and interest in the lives of people in the headlines" (Peterson 1964: 379).[10] The magazine Harrison started, *Confidential*, whose first issue was published in December of 1952, was selling 3.2 million copies per issue by 1955 – more than *TV Guide*, *Reader's Digest*, the *Ladies Home Journal* and the *Saturday Evening Post*.[11] The success of *Confidential* itself is less significant than is the fact that, in the wake of its success, dozens of imitators began publication, initiating a cycle of magazines which flourished between 1952 and 1957 and faded precipitiously thereafter. The total circulation of these magazines, which included such titles as *Behind the Scenes*, *Dare*, *Exposed*, *Hush-Hush*, *Inside Story*, *Lowdown*, *Private Lives*, *Rave*, *Tip-Off*, *Secret*, *Uncensored*, dozens of one-shots and many others, had reached 10 million copies per issue by 1955.[12]

Given its association with celebrity gossip, *Confidential's* links to the Kefauver hearings are not immediately apparent. In part, this association obscures the high portion of content which offered probes of non-celebrity vice, local and regional patterns of corruption, and newly-discovered instances of moral and sexual deviance. *Newsweek*, in 1955, suggested that hearings such as Kefauver's had laid the foundation for the new scandal magazines:

> Having seen more than his share of legitimate scandals and exposures, the reader begins to think that every story must have some kind of "lowdown" beneath its surface, some "uncensored" facts known only to a "confidential" few.[13]

In fact, in the almost endless proliferation of exposé stories, through the large-sized magazines such as *Confidential*, but more markedly in the digest publications which multiplied in the mid-1950s, there is little sense of uncovering the underside of prominent social sites. As in the films discussed above, there is, rather, an emphasis on peripheral and obscure locations for vice, as if it must be found anew each time it is written about and thus sought further and further afield.[14] Again, rather than producing the sense of a nationally all-pervasive criminal conspiracy, these stories, in their seriality, conjure up interminable local particularisms and varieties. What distinguishes these magazines from earlier progenitors, such as the *Police Gazette*, is their emphasis on local or regional patterns of vice or corruption over the individual, sensational crime. This is one result of a production process which did not require journalists on location, but, rather, involved the reshuffling by a limited number of office personnel of less time-bound information and images.

The scandal magazines which flourished from 1952 to 1957 were the object of a moral panic not unlike that directed at comic books during the period. By 1955, associations of druggists had organized a boycott of scandal magazines, and the Post Office had demanded prior approval of issues of *Confidential* before delivering them through the mails. More damagingly, by 1957 the accumulated impact of lawsuits directed at the major magazines by a variety of public personalities had drained their resources and, in some cases, resulted in the imposition of fines. In 1957, after a major libel and obscenity trial in California ended in a hung jury, Robert Harrison, the publisher of *Confidential*, announced that he would "eliminate exposé stories on the private lives of celebrities." Shortly thereafter, he sold the magazine, which continued in a much milder form until the 1970s. Dozens of other scandal magazines died in 1956 and 1957 or began decreasing their frequency until ceasing publication near the turn of the decade. (The year 1957 is, as well, the year of *The Sweet Smell of Success*, a film seen as signalling the destruction of the image of Walter Winchell, and of the sort of urban exposé journalism with which he was associated.)[15]

The cultural form most directly and demonstrably threatened by the new scandal magazines was the Hollywood fan magazine, which had thrived for almost a half century on the basis of intimate connections to the movie studios. (The 1956 Republic film, *Scandal, Inc.* is partly about these changes, though its own exploitation of the scandal magazine fad works against its "critique" of these magazines for instilling an atmosphere of dread in Hollywood.) In the 1950s, the circulation of fan magazines declined significantly, and the form would survive principally through a new association with popular music and a targeting of younger audiences. The scandal magazines themselves had built their success on a tenuous coalition of readers which began to fragment by the late 1950s. Many of the digest-sized magazines, such as *Quick* or *Dare*, overlapped with the cultural traditions and audiences of pin-up culture, a formation now fractured through the appeal of the new, slick men's magazines. More generally, the readership of scandal magazines in the 1950s had encompassed both males and females in roughly equitable numbers. By 1960, the gendering of scandal magazines had become much more pronounced. Exposés of situational (as opposed to personal) vice were more and more the province of quasi-militaristic adventure magazines directed at men. An emphasis on celebrity scandal and the unusual began to configure around the supermarket tabloid and those forms which fed into it.

The line of cultural descent traced here marks the dispersion of exploitational culture, from the first of the Lait and Mortimer books, with their focus on the most monumental of US cities, through the moment of the Kefauver Committee, which shifted much of the attention to mid-sized, underdeveloped cities of the south and midwest. In the scandal magazines which follow in the Committee's wake, there is a further scattering of emphasis, towards a more generalized idea of vice having a location and towards the infinite repeatability of that locatedness.

Economic logic will lead to the proliferation of exposé magazines and films, and these will locate vice and corruption in increasingly minor locales. This drift, combined with the shrinking economic basis for both forms, will enhance their obscurity and marginality. Arguably, the signs of their own degradation will become one source of the fetishistic connoisseurship which has formed around these cultural artefacts.

Geoffrey O'Brien has noted a shift in crime fiction during the 1950s, one marked by a decline in conventional mystery structures and recurrent hero figures as the points of continuity between texts. (The notable exception here is Mickey Spillane's Mike Hammer; O'Brien 1981: 119). The typical paperback crime novels of the 1950s offer a dispersed geography of obscure locations, aberrant characterological structures and themes which have little to do with prominent socio-political questions of the time. This is the case, as well, with dozens of low-budget crime films produced during the decade, films marked by the obscurity of peripheral locations, compressed, situational narratives and a disengagement from the grand social thematics of either the classical gangster film or the postwar *film noir*.[16] One can note, as well, changes in the American comic book, which, in the 1950s, sees a significant decline in recurrent heroes and the proliferation of titles offering unconnected, often blatantly exploitational mini-narratives (Benton 1993).

THE INEXHAUSTIBLE CORPUS

In describing the broad intertextual space described here, the aesthetic category of the *lurid* holds particular pertinence. Despite its multiple and even contradictory historical meanings, the term "lurid" has come, with time, to suggest the textual rendering of sensation. "Lurid" would be used, in the 1930s, to describe the exaggerated histrionics of the pulp magazine cover model or the over-wrought musical punctuation used in movie serials. Applied to a popular culture of the 1950s, its meanings begin to shift, to cluster around a distinctive representation of the jazzy, nocturnal, vice-ridden American city. (For one description of this shift, see O'Brien 1981). The lurid will come to suggest what, in stylistic terms, might be called an angularity: the use of visual and aural figures which beckon into a textual depth. Here a thematics of vice and secrecy are joined to the sorts of seductive promise which typically characterize the exploitation film. Both are given textual form through a visual language inherited from *film noir* and a musical sensibility drawn from small-group jazz, but the respectability of each of these influences is sacrificed. The diagonal rendering of neon-lit city streets, images of doorways or alleys with human figures posed alluringly within, the musical nocturnes common within films of the period – all of these have become the stereotypical markers of a certain kind of popular culture of the 1950s.

Many of these features predate the 1950s and will persist beyond that decade. By saturating a broad section of a particular body of popular culture during this period, however, they have come to define a historically specific imagery of the

American city. This imagery departs from the depictions of the cityscape which are characteristic of films of the 1940s, and particularly of their opening scenes. There, the city is typically busy and chaotic, its skyline monumental and triumphant. In those semi-documentary films of the mid-to-late 1940s which purport to investigate the condition of cities, the characteristic opening is one which establishes an institutional voice and vantage point of authority from which the investigation is undertaken. Normally, in such films, martial-like music accompanies establishing shots of official, monumental institutions.

In the period being discussed here, dozens of films linked to an idea of urban exposé open in ways which promise and withhold an illicit uncovering of secrets. In this, the consolidation of a vocabulary for suggesting the lurid is inseparable from changes in certain textual and paratextual features of American films, changes which help to signal the dissolution of classical Hollywood. These features include the withholding of credits until a locale or situation has been established, the decline of studio orchestras and their replacement by jazz ensembles, and the frequency of on-location footage. All of these have antecedents which precede the 1950s, of course (see Krutnik, this volume), but together they signal a departure from the more ceremonial function of credits in the 1940s, when they framed and announced texts in a more obviously distinct fashion. In such crime thrillers of the 1950s as *Finger Man* (1955), *I Cover the Underworld* (1955), *Private Hell 36* (1954), *While the City Sleeps* (1956), and *The Human Jungle* (1954), opening images of the city build anticipation for moments of imminent sensation. In other genres of the time, in contrast, one sees the growing ornamentalization of credit sequences, their expansion as relatively discrete moments of graphic, dramatic and often musical expression (as in the comedy films of Frank Tashlin, for example). As the lurid exposé film itself begins to disappear, late in the 1950s, we see the rise of a set of stylistic features which signal a new modernity in Hollywood cinema. The alluring depth of exposé credit sequences will give way, as the decade concludes, to the flat, geometrical graphic forms then fashionable in credit sequence design, just as the rise of the caper film will serve partially to divert the crime narrative from a generalized moral panic over urban vice and corruption and in the direction of more mannerist, abstract narrative games.[17]

Paul Willemen has described the cinephilia surrounding much of postwar American cinema as founded on a quest for the fragmentary image seen to bear the marks of an authorial subjectivity. This is, he suggests, an image typically found in the most glaringly inept of texts. "Cinephilia," Willemen writes, "was founded on a theory of the sublime moment, the breathtaking fragment which suddenly and momentarily bore witness to the presence and force of desire in the midst of appallingly routinized and oppressive conditions of production" (Willemen 1980: 3). In the genesis of cinematic auteurism, of course, these fragmentary signs of desire are the marks of the creative individual whose isolation and identification fuel critical analysis (e.g. Sarris 1968). Through the uncovering of this subjectivity, the work may be salvaged, removed from its often degrading conditions of production and generic affiliations and placed within other, more reputable, genealogies.

One of these genealogies, of course, has been that of *film noir*, a category of film whose mapping has been intimately bound up with the elaboration of an auteurist criticism. It seems clear, nevertheless, that the cinephilia which surrounds the 1950s urban thriller is driven only partially by the need to undertake the salvage operation just described. Part of *film noir*'s beauty as an object, Marc Vernet suggests, is that there is always another film to discover (Vernet 1993: 1). This inexhaustibility is true of such categories of film as the B Western, as well, but for the cinephilia which surrounds the urban thriller of the 1950s, the unearthing of new titles has a more obviously illicit dimension. Over time, the archaeological discovery of new, obscure films has become inseparable from an unearthing of the lurid and degraded, as the fan/historian is led into more and more peripheral corners of an industry undergoing dissolution and change. The films still to be turned up will almost certainly be those which come from mid-decade or later, long after links to a postwar disillusionment or to the nobility of the private eye tradition may be made with a straight face. The works which remain are likely, as well, to be residues of the transition which leads from the breakup of the old Hollywood to the emergence of new spaces of filmic exploitation and industrial marginality.[18] One form of cinephilia, of course, has been drawn to such obscurity, to the collecting of examples which not only add to a list but lead towards the boundaries of an industry/system. If the exposé films of the mid-to-late 1950s are a privileged object of this cinephilia, it is because the movement towards industrial marginality is doubled in the discovery of films whose locations are increasingly peripheral and whose ties to Hollywood's standards of propriety and convention are stretched further and further.

ACKNOWLEDGEMENTS

I would like to thank Gordon Harrison, Anne Carruthers and Aurora Wallace who, at various points, have provided invaluable research assistance on the larger project of which this is a part. I am extremely grateful to Don Wallace, for his expertise concerning the popular literature of this period and for his generosity in sharing it with me.

NOTES

1   For example, the poster for *Portland Expose* (1957) reads "*Blistering* in the newspaper headlines! *Nakedly* shocking on the screen! ... *Life* exposed the Portland Story". That for *The Phenix City Story* (1955) suggests that "If *Life*, *Look*, *Sat. Eve. Post*, *Newsweek* and *Time* hadn't exposed this shocking story ... You wouldn't believe it." The opening credits for both *The Phenix City Story* and *Chicago Confidential* (1957) contain graphic representations of the magazines or books from which their "story" is taken.

2   See, for a discussion of one such cycle, Russell Nye's account of the late nineteenth century "inside" city exposé books by Lippard, Judson and others (Nye 1971: 30–1).

3   *New York Confidential* (Lait and Mortimer 1948). This was followed by *Chicago Confidential* (1950), *Washington Confidential* (1951) and *U.S.A. Confidential* (1952). Jack Lait died in 1954. In 1956, Lee Mortimer published *Around the World Confidential* and, in 1960, *Women Confidential*. In the introduction to the latter, he complained of those who had cashed in on the "confidential" formula and "who confused the disclosure of public dirt with the dishing of private dirt" (Mortimer 1960: 12). In 1962, Lee Mortimer published *Washington Confidential Today*, an updated version of the earlier book.

4   For one among many profiles of Lait and Mortimer published following the success of their first co-authored book, see "Hustling Hearsting," *Time* August 30, 1948, p. 48.

5   Among the many lawsuits against Lait and Mortimer were one by a Republican Senator from Maine whom the two had accused of communist sympathies, and another by the employees of a Neiman–Marcus store in Dallas, accused by Lait and Mortimer of being either prostitutes or homosexuals. Among the many accounts of these lawsuits, see "Margaret Smith Wins Retraction: Senator Gets $15,000 From Authors and Publishers of 'U.S.A. Confidential,'" *The New York Times*, October 18, 1956, p. 2, and "Retraction Is Won by

Neiman–Marcus," *The New York Times*, May 6, 1955, p. B3.

6   See Gabler (1994).

7   As William R. Taylor suggests, "[t]he desire to find the village within the city" has been a prominent motive in the work of what he calls "Broadway mythologists" such as Damon Runyon (Taylor 1992: xvi).

8   This version of the story is offered by Lait and Mortimer in *U.S.A. Confidential*, who claim that Kefauver approached them immediately after having his eyes opened by *Chicago Confidential*: "We were in the capital doing the groundwork for *Washington Confidential*. He sought us out. He asked dumb questions and we gave him wise answers – yes, it was all true and enough left over for more volumes" (Lait and Mortimer 1951: 253).

9   The most comprehensive history of the Kefauver hearings is Moore (1974).

10   For claims that television broadcasts of the Kefauver hearings inspired Harrison to begin *Confidential* magazine, see, among others, "The Curious Craze for Confidential Magazines," *Newsweek*, July 11, 1955, p. 50 and Govoni 1990: 30.

11   These are the circulation figures reported in "Sin, Sex and Sales," *Newsweek*, March 14, 1955, p. 88

12   This figure is offered in "The Curious Craze for Confidential Magazines," *Newsweek*, July 11, 1955, p. 50.

13   "The Curious Craze for Confidential Magazines," *Newsweek*, July 11, 1955, p. 50.

14   A random sampling of articles which undertake the sort of exposé described here might include the following, from my personal collection: Abramson, "Montreal Confidential," *Photo*, July, 1953, pp. 12–19; "Cleveland: Den of Vice," *Pic*, March, 1955, pp. 102–5; "Miami's Torrid Night Life," *Tab*, May, 1952, pp. 109–13; "The Rackets are Ruining Cleveland," *Photo*, September, 1952, pp. 3–12; "San Pedro, Port of Sin," *Photo*, June, 1952, pp. 3–13; "Special city-by-city report on 'Sex U.S.A.'," *Pose!*, June, 1955; "Houston – No. 1 Murder City," *Focus*, April, 1953, pp. 36–8; "City Saved By Sin" (on Delano, California), *Pose!*, April, 1955, pp. 30–2; and "Miami – America's Liveliest Port of Entry for Sin," *Bold*, June, 1961, pp. 58–62.

15   The cyclical decline of the scandal magazine, under the effect of retail boycotts and lawsuits, is covered in great detail in the popular press of the time. See, in particular, "Jungle," *Newsweek*, September 17, 1957, p. 102; "The 'Exhausting' Juror," *Newsweek*, October 14, 1957, p. 74; "Confidential Clean-Up," *Newsweek*, November 25, 1957, p. 81; "Confidentially," *The Reporter*, November 3, 1955, pp. 4–6.

16   Examples of these films, which share with the urban exposé film the industrial and generic obscurity discussed here, but not the latter's thematics, would include *Plunder Road* (1957), *Loophole* (1954), *The Big Bluff* (1955), *Fright* (1955), *Highway Dragnet* (1954), *I Cover the Underworld* (1955), and countless others.

17   It might be argued that the form of the caper film is itself geometrical and, in that sense, abstract, through its reliance on the mapping of character relationships and the establishment of a set of narrative possibilities whose success or failure is contingent on causal relationships laid out at the outset. *The Killing* (1956) clearly anticipates the shifts described here.

18   For an insightful analysis of the scholarly dispositions which have taken shape around "trash culture," see Sconce (1995).

REFERENCES

Benton, M. (1993) *Crime Comics: The Illustrated History*, Dallas, Tex.: Taylor Publishing.

Derr, J. (1986) " 'The biggest show on earth': the Kefauver Crime Commission hearings," *The Maryland Historian* 17 (2): 19–37.

Gabler, N. (1994) *Winchell: Gossip, Power and the Culture of Celebrity*, New York: Alfred A. Knopf.

Govoni, S. (1990) "Now it can be told," *American Film* 15 (February): 28–33.

Lait, J. and Mortimer, L. (1948) *New York Confidential*, New York: Crown.

Lait, J. and Mortimer, L. (1950) *Chicago Confidential*, New York: Crown.

Lait, J. and Mortimer, L. (1951) *Washington Confidential*, New York: Crown.

Lait, J. and Mortimer, L. (1952) *U.S.A. Confidential*, New York: Crown.

McGivern, W. P. (1956) *Police Special ("including Three Complete Novels: Rogue Cop, The Seven File, The Darkest Hour")* New York: Dodd, Mead and Company.

Moore, W. H. (1974) *The Kefauver Committee and the Politics of Crime 1950–1952*, Columbia: University of Missouri Press.

Mortimer, L. (1951) "Underworld confidential: will Kefauver be president?" *American Mercury* 72 (329): 515–24.

Mortimer, L. (1956) *Around the World Confidential*, New York: Putnam.

Mortimer, L. (1960) *Women Confidential*, New York: Julian Messner.

Mortimer, L. (1960) *Washington Confidential Today*, New York: Paperback Library.

Nord, D. E. (1991) "The urban peripatetic: spectator, streetwalker, woman writer," *Nineteenth Century Literature* 46, 351–75.

Nye, R. (1971) *The Unembarrassed Muse: The Popular Arts in America*, New York: The Dial Press.

O'Brien, G. (1981) *Hardboiled America: The Lurid Years of Paperbacks*, New York: Van Nostrand Reinhold.

Peterson, T. (1964) *Magazines in the Twentieth Century*, Urbana: University of Illinois Press.

Sarris, A. (1968) *The American Cinema: Directors and Directions 1929–1968*, New York: E. P. Dutton.

Sconce, J. (1995) "'Trashing the academy': taste, excess and an emerging politics of cinematic style," *Screen* 36 (4): 371–93.

Taylor, W. R. (1992) *In Pursuit of Gotham: Culture and Commerce in New York*, New York and Oxford: Oxford University Press.

US Congress, Senate (1951) Special Committee to Investigate Organized Crime in Interstate Commerce, *The Kefauver Commitee Report on Organized Crime*, New York: Didier.

Vernet, M. (1993) "*Film noir* on the edge of doom," in J. Copjec (ed.) *Shades of Noir: A Reader*, London: Verso, 1–31.

Willemen, P. (1980) *The Edinburgh Film Festival and Joseph H. Lewis* (program notes) Edinburgh: Edinburgh Film Festival.

6

CINÉCITIES IN THE SIXTIES
•

Antony Easthope

Unreal city
T. S. ELIOT, *The Waste Land*, 1922

A classic trope in American *film noir* is an opening in which a voice warns us of the darkness and danger of the city (over shots of skyscrapers at night with lighted windows) and then promises to show us that darkness. Lynch's *Blue Velvet* (1986) similarly sets up an opposition between light and darkness, surface and depth, between the bright and cheery suburban streets of daytime Lumberton and another night-time Lumberton, hidden beneath (or inside) the first. Yet the film establishes the opposition only to breach it. Sinister details from the other Lumberton constantly intrude into the innocent suburban spaces (as Jeffrey walks to visit the police detective along leafy streets, the trees suddenly darken to become full of menace), while the good, upright people of Lumberton are shown to be subject to desire for what the other side of the town has to offer. Unlike the classic gangster movie, *Blue Velvet* stops us discriminating firmly between the apparent and the real, surface and depth, light and dark. Rather, we are encouraged to think of the two moral dimensions of the city as present simultaneously in the same physical space. A lecture delivered by Michel Foucault in 1967 anticipates something of the effect envisaged by *Blue Velvet*.

In his talk Foucault speculated that Western culture was undergoing a change in experience of the structuring of time and space, and their relation:

> The great obsession of the nineteenth century was, as we know, history: with its themes of development and of suspension, of crisis and cycle, themes of the ever-accumulating past, with its great preponderance of dead men and the menacing glaciation of the world . . . The present epoch will perhaps be above all the epoch of space. We are in the epoch of simultaneity: we are in the epoch of juxtaposition, the epoch of the near and far, of the side-by-side, of the dispersed.[1]

On this basis Foucault theorises the possibility of the heterotopia: 'The heterotopia is capable of juxtaposing in a single real place several spaces, several sites that are in themselves incompatible.'[2]

I am not the first to find these speculations very suggestive at the same time as I respond to them with a certain scepticism. Foucault is asserting that a classic or traditional concern with history and temporality persists across modernism until . . . until when? Foucault's aposiopesis becomes retrospectively significant. The lecture was given in 1967, and so 'the present epoch' he envisages in which the transformation of space will lead to a 'juxtaposing . . . [of] several sites that are in themselves incompatible' looks forward to transformations beyond modernism that we have since learned to name, live with and live into. From the heart of the modernist 1960s an as yet unnamed theme announces itself: postmodernism. It is this shifting in the terrain, the sense of spatialisation, or rather the *relation* between temporality and spatialisation, which I aim to trace in some films from the 1960s, texts which we might now see as solidly pre-postmodern. First, though, there are some questions regarding space, cities and subjectivity.

CLASSICAL SPACE: VISION AND SUBJECTIVITY

Foucault speaks confidently of space as potentially heterotopic because he assumes without question that what has been termed *classical* space is no longer available. In a famous paragraph in *The Production of Space* Henri Lefebvre outlines what that sense of space implied while confirming its loss in the period of Modernism:

> The fact is that around 1910 a certain space was shattered. It was the space of common sense, of knowledge (*savoir*), of social practice, of political power, a space thitherto enshrined in everyday discourse, just as in abstract thought, as the environment of and channel for communications; the space, too, of classical perspective and geometry, developed from the Renaissance onwards on the basis of the Greek tradition (Euclid, logic) and bodied forth in Western art and philosophy, as in the form of the city and town.[3]

Our relation to space, including city space, is epistemological. And since subject and object are always produced together, the quality and design of city space will pose reciprocally a certain position and definition for the subject. Although alert to the subjective effectivity of classical space, Lefebvre does not theorise the form in which a coherent, fixed, knowable space – classical space – offers a position to the subject. For a preliminary account of that structuring I shall turn to Lacan's ˙lysis of subjectivity and vision.

tells the story of a small boy on a fishing trip who pointed to a floating ˙d said, 'You see that can? Do you see it? Well, it doesn't see you.'[4] ˙e me but *someone* there might, since whatever I can see consti- ˙ich I could be seen: 'I see only from one point, but in my ˙ed at from all sides.'[5] Maintaining itself in relation to sight, ˙ itself, presuming that it sees the Other when it always appears

CINÉCITIES IN THE SIXTIES 131

at a point within the Other, for the Other. Subject, then, is positioned by object but denies or disavows it is an effect of that positioning.

At the Renaissance a number of developments in the organisation and representation of space work to privilege the position of the I as transcendent and a point of origin by denying awareness of its construction. Unprecedented priority is given to the faculty of vision, a whole series of new practices inciting the subject to see the world as a picture, an object for knowledge and visual pleasure. Such scopophilia is widely promoted by the introduction into painting of monocular perspective. This, as Lacan argues, is able to efface the gaze of the Other by rendering it as the vanishing point of the perspective, not a point from which I could be seen but rather as just part of an object for which I am a subject.[6]

At the same time as a regime of visual representation encourages the subject to look at the world as a picture, that world is constructed to appear as a picture. Newly built Renaissance and post-Renaissance cities are structured according to the logic of grid design and geometrically unified space (Euclidean space), those principles informing both the architecture of individual buildings and the design of cities. Perspectives inscribed in stone answer to the expectations of monocular perspective.

Classical space, therefore, affords the subject a position which is seemingly coherent, unified and self-sufficient, a Cartesian subject. But it does so only on condition that the dialectical relation between subject and object which produces a position for the subject is concealed, effaced, denied.

This is also the case with representations of classical space, notably the Quattrocento tradition in painting. In this signifying practice the subject's autonomy is ensured by the visual object's apparent coherence and fixity 'out there'. Again, the requirement for the effect is that the means of representation itself be held in the background, away from attention. Establishment of classical space in representation depends on an effect of transparency, a foundational convention carried through into the 'narrative space' of mainstream cinema. In considering some 1960s movies I shall be alert for any uncertainty about the capacity of cinema to *represent* because that is a symptom of the instability of classical space; it also implies that the sovereignty of the Cartesian subject is in doubt.

On this basis I shall explore three attitudes toward representing the city and its spaces in 1960s Western cinema:

1 The city is 'just there', naturalised;
2 A celebratory and utopian presentation of the city;
3 The city as sign and realisation of dystopia.

I shall have most to say about the third.

THE NATURALISED CITY

During the 1930s the insistence of the city was strongly marked. This is apparent in European examples, notably Lang's *Metropolis* of 1926, but it is also clear in American cinema, particularly the gangster movie, in examples from the early Depression years, such as *Little Caesar* (which initiates the genre in 1930), *Public Enemy* (1931) and *Scarface* (1932). Of the city background for this genre Colin McArthur writes:

> The sub-milieux of the gangster film/thriller are, in fact, recurrent selections from real city locales: dark streets, dingy rooming-houses and office blocks, bars, night-clubs, penthouse apartments, precinct stations and, especially in the thriller, luxurious mansions. These milieux, charged with the tension of the violence and mystery enacted within them, are most often seen at night, lit by feeble street lights or more garish neon signs.[7]

Communication is by telephone, lighting is of course electric, transport is epitomised in the key figure of the modern city, the motor car. Such films show the city as in question, novel. It is so also on the other side of the world, for example in Vertov's Soviet experimental film, *Man with a Movie Camera* (1929), which is held together round a narrative of 'a day in the life of the city'.

In most films of the 1960s the city is simply unmarked; it is just *there*, a normal backdrop, invisible as such, the presumed condition of modernity, 'the world of all of us', in which (as Wordsworth said) 'we find our happiness, or not at all'.[8] However, in this decade, when the city *is* marked as city, it features as either utopian or dystopian. It should be noted that both modes, utopian and dystopian, are dominated by a sense of temporality and history, arising from the assumption that we are moving towards a world which is either much better or much worse. Either way, the conditions for that future are already existent in the present moment (each mode, in fact, is really a way of commenting on the present).

UTOPIAN CITIES

In the first James Bond film, *Dr No*, which came out in 1962, Bond meets a woman in a high-class casino, and goes home with her only to discover that he has an assignment that forces him to fly to Jamaica the next morning. The film ·ts from suggestive fore-play to a jet aircraft landing at an airport, then to everyday ·s inside the airport terminal, Bond looking for a taxi, making a phone call ·driven off. This sequence may stand for dozens of 1960s movies in ·tory take on the city. Typically, we are shown the hero in his ·hots around the terminal, a plane takes off and lands at a ·odern airport with a very similar terminal. And this kind of occurs whether the hero is travelling to smash Smersh (in the cycle of spy-thrillers) or going to Los Angeles to see his girl (as in romance movies such as *The Graduate*, 1967).

Why do we see so much of jet aircraft (never propeller) landing and taking off, taking off and landing, accompanied always by the little puff of smoke and screech when the tyres first hit the tarmac, the symbolic vehicle of the 1960s as much as the car was for the 1930s? Working in excess of what is needed to explain narrative, these repeated airport sequences serve to glorify a utopian vision of intercity technology, a fulfilment of a nineteenth century utopian dream. Science enables a man to move through global city-space as though he were time-travelling; arrival at his destination will present him with a spectacle both exotic and erotic which he will master without difficulty (the concentration is masculinist, the sexual undertone always hinted at, as when in *Dr No* Bond making love with the woman cuts to the aircraft landing). The optimistic implication is that while some live like that now, in a foreseeable future all will be able to live like that. Dystopian presentation of the city is less common and much more interesting.

DYSTOPIAN CITIES

In Antonioni's *L'Avventura* (1960) a group of rich people go on holiday in the Sicilian islands; one simply disappears and two of the others go to look for her (and begin to fall in love with each other). At one point in their quest, while driving, they come to a newly built and apparently deserted town (Plate 6.1). The sequence gives an extraordinary representation of the city, one which owes much to the paintings of de Chirico and also to Antonioni's own former career as an architect. Claudia (Monica Vitti) and Sandro (Gabriele Ferzetti) enter an empty and blank white-walled cityscape which does not result from centuries of casual and 'organic' accretion, street by street, but is manifestly a planned construct, lacking any quality of the given, set down in an equally arid natural landscape or waste land (we wonder if the buildings are for a cemetery).

Here the city is a site of alienation – less social alienation than a full-blown transcendental alienation to be understood as the view that formerly human life in cities was real, a natural inheritance, but now it is made up, a merely arbitrary construction from which something transcendental is felt to be missing. When Claudia goes to a darkened window, calls 'C'e nessuno?' ('Is no one there?') and is answered by the echo of her own voice from the empty interior she only too obviously symbolises man (woman) without God. Confronted with this perceived absence, the couple drive on and, in the very next scene, try to fill the void by making love in the middle of the dry landscape. Sandro appears bored, for him this is normality, but Claudia is deeply struck. For her the white perspective of the vacant city has been a cause for anxiety.

The sequence invokes classical space via a dystopian present defined by a sense of historical past, through temporality – what was thought to be there is there no longer and is marked by its absence. But one shot at the close of the sequence

6.1 *L'Avventura*
Courtesy of BFI Stills, Posters and Designs.

opens on to other possibilities: as Sandro and Claudia prepare to drive off, viewed from a sidestreet, the camera tracks mysteriously towards them. I would want to think of this strange and unnerving shot in two contexts. According to firm cinematic tradition the camera moving towards a person, unnoticed by him or her, signifies a point-of-view shot and therefore an observer (as in the *Halloween* series, frequently accompanied by the sound of disturbed breathing). If this is the case, we might feel that some sinister presence of the deserted city is watching the couple. Alternatively, this otherwise unmotivated tracking shot simply calls attention to the presence of the camera and so the cinematic means of representation. In this case, we are invited to link alienation in the city with a certain erosion of classical cinematic space.[9]

Alphaville (1965) displays Godard's wonderful dystopia in which Lemmy Caution (a kind of *film noir* detective) comes to the city of Alphaville on a commission (Plate 6.2). Run by a computer with a nauseatingly distorted voice, Godard's sci-fi future-city appears wholly predictable, scientifically determined, the exercise of ˜otalising rationality which extends to food, travel, work, sexual relations, architecture, and details of design.

˙*lle* the settings are all in fact from contemporary 1965 Paris, none is ˡ so the text's meaning is articulated in terms of historical tempo- ˙organic' and 'human' past (when though?) adduced to attack pearing in the present. Two readings of 'Alphaville' suggest the text represents the centralised undemocratic hierarchy of

6.2   *Alphaville: Une Etrange Aventure de Lemmy Caution*
Courtesy of BFI Stills, Posters and Designs.

corporate capitalism – in which case it makes a political assertion. Or this demonstration of the contemporary city as alienating and inhuman performs a condemnatory statement from a position of humanist transcendentalism, one comparable (say) to that in Orwell's *Nineteen Eighty-Four*.

Either reading – political satire or humanist denunciation – relies on a coherent viewpoint towards modernist urban space. Spatiality in this respect remains stable and knowable, however dystopian, though this residual assurance is, I think, skewed in two directions. One is that the perceived inhumanity of the city is taken as a motive and correlative justification for the intensity of the sexual relation (Lemmy Caution with the Seductress 3rd Class played by Anna Karina). Another, associated with this, is the traditional seduction of the *noir* narrative. These, I think, pull the text away from its dystopian commitment, tending to make us feel rather that this may be simply how it is, now.

The view that *Alphaville* tends towards both a modernist assertion (history lets us judge the city as inhuman) and a postmodernist sceptical acquiescence may be confirmed from another direction. In hindsight, we can now see how much of *Alphaville* is picked up in Ridley Scott's *Blade Runner* (1982). Godard's city of endless night lends *Blade Runner* its stylistic consistency, as well as something of its lighting, use of music, and even acting style. *Blade Runner*, too, sets a *noir* love affair – Deckard and Rachel – at its narrative centre (despite Scott's efforts to reduce this feature by omitting the voice-over in 'The Director's Cut', 1992).

BLOW-UP

Antonioni's *Blow-Up* (1966) is resolutely and uncompromisingly contemporary in its portrayal of Swinging London in the 1960s. David Hemmings plays a David Bailey-type photographer, who is disillusioned and disaffected. Idly, he visits a park, meets a woman (Vanessa Redgrave) who is up to no good, photographs her in the park with a male friend, they disappear (Plate 6.3). Later, he develops the film, enlarges it, he begins to think that the man in the park was murdered, that someone was waiting for him in the bushes. The narrative question is: was the murderer actually there or is he just a paranoiac symptom?

Blow-Up is mostly concerned with nature, not the cityscape, crucially in the scene in the park, though this itself has significance. Before, in *L'Avventura*, Antonioni's new city was surrounded by natural landscape; now nature is a park, landscape surrounded by city. In an extended sequence the camera lingers on the trees and the grass in the park, which appears mysteriously, indefinably alive (the banal sound of the wind through the leaves is augmented on the sound-track to suggest something else, whispering perhaps). The real, it seems, defies representation, resists the hero's penetrating and mastering masculine gaze (unlike, say, the protagonist's assurance with his camera in *Rear Window*, 1954).

Afterwards, in his studio, the photographer makes a series of enlargements of the shots he has taken, searching for a knowledge behind the image, in the image. The truth at stake is a possible murder but when a face and a gun hidden in the bushes finally do emerge the images are undecidable, for they may be just marks on the negative, a product of the enlargement process itself. Representation now constructs what it represents, and there is no fixed object to be known outside the process of representation. Classical space can no longer be relied on, nor, correspondingly, can the autonomous, mastering and implicitly masculinist subject (the photographer fears for his own sanity).

This scene from *Blow-Up* looks back to the moment at the strange, empty town in *L'Avventura* when the camera was no longer concealed (in the way of classic realism) but without narrative motivation moved in on the couple. And it looks forward to *Blade Runner* and a similar effect when Deckard, with the resources of more modern technology, persists in enhancing an image until he reaches the manufacturer's number of an artificial snake-scale. Unlike *Blade Runner* (less post-modern, in my view, than its spectacular *mise-en-scène* promises), *Blow-Up* begins to put the text's own epistemological status in question.

Blow-Up also shows much less confidence in working out either a utopian or a dystopian attitude towards the city. In comparison with *L'Avventura* it has lost much (not all) of its nostalgia for a metaphysical dimension, and the question of

6.3   *Blow-Up*

transcendental alienation hardly enters the movie at all. The film ends with a kind of Fellini-type mime troupe playing tennis without a ball; the imaginary ball goes out of court, near to David Hemmings; he picks it up and throws it back. London as the contemporary city is no longer imaged by contrasting a false present with a fuller past, according to a logic of historical temporality which, arguably, presupposes the knowable coherence of classical space. That sense of space within which the city may be held as the object of knowledge is not so readily available.

If, with postmodernism, 'most people have lost the nostalgia for the lost narrative',[10] modernism is defined by precisely that feeling of nostalgia for what is lost; if postmodernism, according to the cultural logic of late capitalism, is characterised by normalisation of 'a new kind of flatness or depthlessness',[11] modernism was marked by a sense of shock which continues to invoke the possibility of depth and significance endowed by a larger narrative. Modernism, then, is presided over by a sense of loss, however articulated, while postmodernism is not. The films of the 1960s looked at here accept the modernity of the city without remark, celebrate it or see it as the city of dreadful night. Yet whether utopian or dystopian in tendency, they affirm either that the contemporary city completes a historical vision or that it is a place left after the collapse of the Grand Narratives. The townscape that I have singled out from *L'Avventura* seems replete with the death of God, and Godard's film may be read to say that Alphaville would be unlivable for fully human beings accustomed to the kind of community we used to have.

However, in Foucault's less pre-emptive and perhaps more neutral terms, with the transition from modernism to postmodernism, temporality gives way to spatiality, history to simultaneity, juxtaposition and heterotopia. That hypothesis, too, finds support in the texts under review. A movie such as *Blow-Up* begins to imply that several epistemologically incompatible spaces might be juxtaposed simultaneously, and that opposed frames of representation might be represented together. Classical space, still retained into the modernist 1960s, begins to dissolve there: dystopia foreshadows heterotopia.

I would not want to end by shutting up the discussion too firmly. When in 1980 Foucault was interviewed he recalled a response from 1967 to his ideas about the shift from temporality to spatiality. 'At the end of the study', Foucault says, 'someone spoke up – a Sartrean psychologist who firebombed me, saying that *space* is reactionary and capitalist, but *history* and *becoming* are revolutionary'. And he continues: 'This absurd discourse was not at all unusual at the time. Today everyone would be convulsed with laughter at such a pronouncement, but not then.'[12]

I am not sure that historical time can be laughed out of court so easily, especially now in closing an essay whose perspective has been historical from start to finish.

NOTES

1 Michel Foucault, 'Of Other Spaces' [lecture, '*Des Espaces Autres*', 1967], tr. Jay Miskowiec, *Diacritics* 16, 1 (Spring 1986): 22–7; p. 22.

2 *Ibid.*, p. 25.

3 Henri Lefebvre, *The Production of Space* [1974], tr. Donald Nicholson-Smith (Oxford: Blackwell, 1991), p. 25.

4 Jacques Lacan, *The Four Fundamental Concepts of Psycho-Analysis*, tr. Alan Sheridan (London: Hogarth Press, 1977), p. 95.

5 Lacan, *Fundamental Concepts*, p. 72.

6 See Lacan's complex account of the two imagined cones assumed by monocular perspective, *Fundamental Concepts*, pp. 65–119; see also Stephen Heath, 'Narrative Space', in *Questions of Cinema* (London: Macmillan, 1981), pp. 19–75 and Norman Bryson, *Vision and Painting* (London: Macmillan, 1983).

7 Colin McArthur, *Underworld USA* (London: Secker and Warburg/BFI, 1972), pp. 28–9.

8 William Wordsworth, *The Prelude*, Book 11, ll. 143–4.

9 See again Heath, 'Narrative Space'.

10 Jean-François Lyotard, *The Postmodern Condition* (Manchester: Manchester University Press, 1984), p. 41.

11 Fredric Jameson, 'Postmodernism, or the Cultural Logic of Late Capitalism', *New Left Review* 146 (July/August 1984): 53–92; p. 60.

12 Michel Foucault, 'Space, Power and Knowledge' [1984], tr. Christian Huber, in Paul Rabinow (ed.) *The Foucault Reader* (Harmondsworth: Penguin, 1986), 234–56; pp. 252–3.

7

FROM RAMBLE CITY TO THE SCREENING OF THE EYE

Blade Runner, death and symbolic exchange

•

Marcus A. Doel and David B. Clarke

TYRELL WHAT SEEMS TO BE THE PROBLEM?
ROY   DEATH.

(Blade Runner)

FORGET *BLADE RUNNER*

Early in the 21st century, THE TYRELL
CORPORATION advanced robot evolution
into the NEXUS phase – a being virtually
identical to a human – known as a *Replicant.*
The NEXUS 6 *Replicants* were superior
in strength and agility, and at least equal
in intelligence, to the genetic engineers
who created them.
Replicants were used Off-world as
slave labour, in the hazardous exploration and
colonization of other planets.
After a bloody mutiny by a NEXUS 6
combat team in an Off-world colony,
Replicants were declared illegal
on earth – under penalty of death.
Special police squads – BLADE RUNNER
UNITS – had orders to shoot to kill, upon
detection, any trespassing *Replicant.*
This was not called execution.
It was called retirement.

(Blade Runner, opening text)

Beyond the trivial reasoning that a book on the cinematic city is more or less obliged to say something about *Blade Runner* (1982; 'The Director's Cut', 1992), what possible justification could there be for yet another work on this much-laboured film? The most defensible reason for us is quite simply that *Blade Runner* is not what one thinks – particularly if one thinks that it is an exemplary *mirror* of our own postmodern condition (which gives back from a fictive elsewhere the image of a space-time that will have been ours here and now). In our engagement with the film, we will insist that *Blade Runner* is not a mirror but a *screen*. Cinema does not re-present, re-produce, re-play, or re-flect. Hence, as Deleuze (1986, 1989) demonstrates, conceptualization should work 'alongside' rather than 'on' the cinema: resonance rather than reflection; encounter rather than capture; invention rather than re-presentation. In other words, whilst the mirror is always already given over to and territorialized by something other – which invariably turns out to be a repetition of the Same – the screen is always already immanent to itself. Accordingly, we will not attempt to identify the structure, meaning, location and significance of the image in the mirror but rather to resonate productively with the cinematographic conceptualization inherent to *Blade Runner*: our thoughts on the film will have been an externalized flow, set in motion alongside the flow of illuminated celluloid.

None of this is, of course, to deny that *Blade Runner* has already achieved the oxymoronic status of a canonical postmodern cultural artefact.[1] Its relation to the postmodern stems in part from its splicing of different filmic genres, particularly science fiction and *film noir* (McCaffery, 1991; Grist, 1992). But its status as a postmodern classic cannot be accounted for by its cinematic qualities alone. Thus, in addition to the film's hybrid genre and the double coding of its cinematography, *Blade Runner*'s many commentators have remarked upon: its fractal geography; the interruption of temporality; the triumph of flexible accumulation within the hollow husks of global corporations; the fusion of the mechanisms of capital accumulation and governance; the adsorption of referentiality and representation through a proliferation of simulacra and simulations; the lack of authenticity and the indeterminacy of identity; the short-circuiting of memory, genealogy, and history; the omnipresence of the Fourth World; the slow motion catastrophe of space-time decomposition; and the banality and fatality of living on in the hereafter.

In a wider frame, the disjunctive temporality inherent to the generic hybrid of science fiction and *film noir* – a fusion sometimes dubbed 'Tech Noir' after the bar featured in *The Terminator* – has received significant theoretical attention (Penley, 1991). Needless to say a considerable amount of this has certainly focused directly on *Blade Runner* (Plate 7.1); in the diegesis of this film, the Los Angeles of 2019 is nothing other than the almost total realization of the dystopic horrors that remain only virtual today. Hence, the world of *Blade Runner* is not a space-time to come but a space-time that *will have been*. The future is today, and today has always already fled back to the future (im)perfect; these are truly sliding times (Doel, 1992, 1994). In parenthesis, let us note that this aphanisis and retroflex

7.1   Sci-fi meets *film noir*: Harrison Ford as the blade runner, Deckard
© Warner Bros. Courtesy of BFI Stills, Posters and Designs.

punctuation of the present – its ghosting, or hauntology (Derrida, 1994), so to speak – resonates perfectly with the ontological flicker of both subjectivity and moving pictures: motionless (s)trips drifting in p(l)ace. In fact, the film's director, Ridley Scott, has himself described *Blade Runner* as 'a 40 year-old film set forty years in the future' (Kennedy, 1982: 66). And so, 'at once archaic and futuristic' (Alliez and Feher, 1989: 41), *Blade Runner* implies that, somewhere along the line, the circuits of space-time have been crossed. It has deconstructed its entire universe. Yet as if to prevent this short-circuit from electrocuting the future in a veritable ex-termination of all signs of life, *Blade Runner* acts as an untimely Earth, running this apocalyptic dis-charge to ground. So, whilst we might hanker after a lost utopia where the virtual retains its specificity – as that which can always be actualized otherwise – or cling to the sweet dream that the future is yet to come, the dystopic world of *Blade Runner* suggests that these very possibilities will long since have been (retroactively) scrambled and disseminated.

It is in this context that *Blade Runner* has been held up as perhaps *the* exemplary mirror of the postmodern (Bruno, 1987; Harvey, 1989; Wakefield, 1990). But there is something of a supreme irony to all of this. In the very opening sequence of the film, we are presented with the image of a *false* mirror (Plate 7.2), figured as an anonymous, unblinking eye, transfixed on infinity, in which we see the infernal

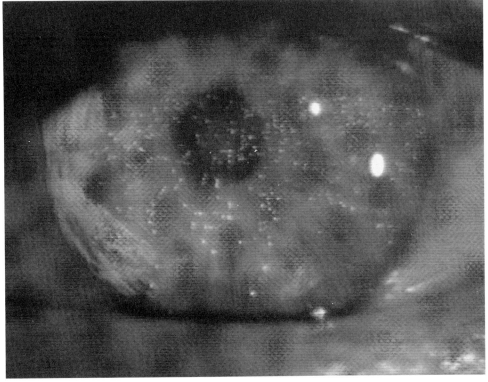

7.2 (a) *Le Faux Miroir* by Magritte
© ADAGP, Paris and DACS, London, 1997. Courtesy of AGK London.

7.2 (b) *Blade Runner*'s 'false mirror'

reflections of an inhuman cityscape – complete with plumes of flame – as if alluding to what Baudrillard (1993a) refers to as 'The Hell of the Same'. And so, where the burgeoning orthodoxy on *Blade Runner* has seen a mirror in which to view the material and social infrastructure, and the socio-political epiphenomena, of a postmodernized capitalist city (Alliez and Feher, 1989; Bruno, 1987; Harvey, 1989), a decentred (in)human subjectivity (Silverman, 1991), and a crisis of representation that serves to short-circuit the teleological metanarratives of History and the real (Marder, 1991), we find a screen that ex-terminates these very problematics themselves – problematics that none the less continue to circulate as so many hauntings and phantomalizations. Initially, therefore, we attempt to explicate how these mirroring-effects break down, in both their originary conceptualization and their supposed exemplary depiction in *Blade Runner*. In their stead, we suggest that the film will have been screened from an altogether different perspective – in accordance with the principle of *symbolic exchange* (Baudrillard, 1993b) – by moving towards a consideration of the economies of death that circulate through(out) the film.

THE FALSE MIRROR: *BLADE RUNNER*, CAPITALISM AND SCHIZOPHRENIA

'Ramble City', Giuliana Bruno's seminal essay on *Blade Runner*, marked the first attempt to elaborate the film as 'a metaphor of the postmodern condition' (Bruno, 1987: 62). Drawing on Fredric Jameson's account of postmodernism as the cultural logic of late capitalism (Jameson, 1983, 1984; see also 1991), the postmodern is implicitly theorized by Bruno as a historically specific mo(ve)ment, colonizing the space-time of a receding modernity. David Harvey's (1989) variation on Bruno's analysis – theorized in relation to a new mode of 'flexible accumulation' deriving from Aglietta's (1979) regulation theory rather than Mandel's (1975) 'late capitalism'; and inflected through the trope of space-time 'compression' rather than space-time 'colonization' (Lefebvre, 1991; Kirsch, 1995) – similarly affirms the idea that postmodernism is but the cultural clothing of a qualitatively different phase of capitalist development. Indeed, 'compression' would appear to be one of the most enduring motifs of the film, in the sense that everything appears to be in absolute proximity; everything seems to be overexposed in a veritable porno-geography of obscene promiscuity.[2] Yet compression can take place if, and only if, something remains exterior to it, which thereby undergoes a corollary de-compression. Compression can never be total and all encompassing; it is always relative, a relation of speed and slowness. Moreover, compression changes nothing – and this is a decisive weakness for any engagement with either *Blade Runner* or postmodernity. Despite the variegated circulation of terms, the terms themselves remain the same. They only ever enter into new arrangements. This is why we prefer Baudrillard's (1994) notion of space-time *implosion* – the volatilization and ex-termination of terms – and Derrida's (1994) notion of space-time *disadjustment*

– the destabilization which is always already *in* the terms, and not merely *around* them (Clarke and Doel, 1994; Doel, 1992, 1994, 1996).

Despite Harvey's forceful reaffirmation of such a base–superstructure relation, it is arguably Bruno's account that has set the parameters for all subsequent discussion of the film – particularly in terms of its spatiality (characterized by the city's architectural pastiche) and temporality (marked by a schizoid experience of time). Underwriting much of the literature, therefore, has been a conceptualization of postmodern*ity* as an epoch progressively instituted in space-time in accordance with a logic of de-differentiation, which explodes the relative autonomy of the distinct spheres of economy, polity, civil society, culture and so on, through the negatory process of postmodern*ization*; and from which a correlative cultural logic or 'structure of feeling' (postmodern*ism*) may be directly discerned. *Blade Runner* may then be viewed as an exemplary mirror of the essential, latent and virtual horrors of our own postmodern condition ramified to the nth degree – in which the fractal stage of value reigns supreme and the cultural logic of everyday life is triangulated by banality, fatality and simulation (Clarke and Doel, 1994; Doel and Clarke, forthcoming). Whilst we do not necessarily dissent from such an understanding of the fissiparous present, we see none of this exemplified in the film itself.

Mike Davis (1992: 1), too, remains incredulous to what he describes as 'LA's own dystopic alter ego'. Noting that whilst 'Ridley Scott's particular "gigantesque caricature" may capture ethno-centric anxieties about poly-glottism run amuck', Davis (1992: 2) suggests that 'it fails to imaginatively engage the real Los Angeles landscape – especially the great unbroken plains of ageing bungalows, dingbats and ranch-style houses – as it socially and physically erodes into the 21st century.' For Davis (*ibid.*), the film 'remains yet another edition of [a] core modernist vision – alternately utopia or dystopia, *ville radieuse* or Gotham City[3] – of the future metropolis as Monster Manhattan.' And so Davis yawns at the way in which the film stages the future 'as a grotesque, Wellsian magnification of technology and architecture', preferring to conjure a 'Gibsonian' map of the future, which unfolds through delicate extrapolation rather than crude magnification (*ibid.*).[4] Nevertheless, whilst eschewing the obsession with dystopic gigantism in favour of careful extrapolation, Davis persists in viewing *Blade Runner* as an attempt to mirror the essential, latent and virtual Los Angeles of today. At bottom, he tells us to forget *Blade Runner* in so far as it is a mirror with distortions: when we refuse to be seduced by 'the mile-high neo-Mayan pyramid of the Tyrell Corp. [which] drips acid-rain on the mongrel masses in the teeming Ginza far below', and we remove 'the overlays of "Yellow Peril" . . . and "Noir" . . . as well as a lot of high-tech plumbing retrofited to street-level urban decay, what remains is recognizably the same vista of urban gigantism that Fritz Lang celebrated in *Metropolis*' (Davis, 1992: 1–2).[5] Hence, Davis persists in conceiving of *Blade Runner* as a mirror in which to find reflected the alternative futures of Los Angeles, failing to make a decisive move from mirror to screen. Before summoning such a move ourselves, however, it is worth summarizing those features deployed in *Blade*

Runner that have been marshalled in the reading of the film as an exemplary mirror of the postmodern. We wish to argue that, whilst the film *appears* to reflect something of capitalism, subjectivity, history, and the real, these are simply so many false mirrors – so many hauntings or spectral manifestations.

For Bruno (1987), *Blade Runner* provides a powerful vision of the postindustrial city: as the city in ruins. On this account, the relation between postmodernism and late capitalism is evident particularly in the film's representation of postindustrial decay, the proliferation of waste having come to serve as an index of the accelerated turnover time of a new phase of capitalism. This, in turn, is reflected in an aesthetic of pastiche – evidenced, for instance, in the costumes of the *Replicants* Pris and Zhora, where 'consumerism, waste, and recycling meet' (*ibid.*: 64; Plate 7.3). However, 'It is in the architectural layout of *Blade Runner* that pastiche is most dramatically visible and where the connection of postmodernism to post-industrialism is evident' (*ibid.*: 62) – pastiche being, for Jameson (1983: 113), the aesthetic form that articulates 'the postmodernist experience of space'. Accordingly, the film portrays the city as 'a synthesis of mental architectures, of *topoi*' (Bruno, 1987: 66), which provides an excess of scenography; references that are simply there, for no apparent purpose (save, perhaps, for a game of recognition on the part of the spectator: Umberto Eco's 'double coding').[6] Thus, whilst 'The city is called Los Angeles . . . it is an LA that looks very much like New York, Hong Kong or Tokyo' (*ibid.*: 65–6).[7]

Drawing on Laporte (1978; see also Thompson, 1979), Bruno implies that waste attains a new specificity within the postmodern city – as the return of everything that was hurled outside of modernity's performative criteria of order, efficiency, purity and stability. Whereas the modern city was manifestly based upon a hierarchical logic of binary opposition (an ordering principle predicated on the repression of difference and the subordination of heterogeneity to a single, positively valorized term; cf. McArthur, this volume), the postmodern city is witness to the flattening of such hierarchies and the implosion of such oppositions. 'The postmodern aesthetic of *Blade Runner* is thus the result of recycling, fusion of levels, discontinuous signifiers, explosion of boundaries, and erosion' (Bruno, 1987: 65). Much of the city appears to be in a state of advanced decay, paralleled in the 'accelerated decrepitude' or premature ageing of J. F. Sebastian, the genetic engineer through whom the renegade *Replicants* seek to make contact with Tyrell, their maker. But whilst the urban backdrop of *Blade Runner* reeks of decay, the marked contrast between J. F. Sebastian's apartment (in the otherwise deserted shell of LA's Bradbury Building) and the high-tech pyramidal structure that houses the Tyrell Corporation speaks of more than simple urban deindustrialization. The archetypically hollow Tyrell Corporation implies monopoly capitalism disseminated to the nth degree through subcontracting and outworking; whilst the explosion of a Fourth World underclass in the interstices of the city speaks of the hyper-deskilling of labour – to the extent that even a genetic engineer such as Sebastian is forced to seek residence in a decaying shell. Indeed, the spatialization of *Blade Runner*'s social relations are characteristically heterotopic, permitting waste and decay to be

7.3   Zhora flees through the city streets
© Warner Bros. Courtesy of BFI Stills, Posters and Designs.

juxtaposed with the architectural splendour and plush interior of the Tyrell headquarters. However, as Culler (1988: 182) remarks, any sense of a distinctively postmodern 'return of the repressed' with respect to waste is mistaken: it is 'not that
the economic system has brought the postmodern world an increase of rubbish,
and that art has participated in this, but that the element of rubbish . . . [has] long
been a part of sign systems and systems of value all along.' Or again: 'we cannot
dispose of rubbish with a narrative about its emergence or new role in the postmodern world but must reflect on the structure that locates this sludge or dross
within or at the heart of systems of value or language' (*ibid.*).[8] Clearly, then, there
is something unconvincing about the equation of *Blade Runner*'s postindustrial
urban pastiche with the postmodern.

Without further ado, therefore, let us turn to *Blade Runner*'s treatment of temporality, and in particular its relation to the real. According to Bruno (1987: 67),
'The narrative "invention" of the replicant is almost a literalization of Baudrillard's
theory of postmodernism as the age of simulacra and simulation.'[9] Sadly, however,
there has been considerable confusion and misunderstanding on this matter. For

example, most commentators assume that the *Replicants* want to become human; that they desire a *human* lifespan and subjectivity. To some extent this is supported by Roy Batty's characterization of the *Replicant*'s existence in terms of labour and slavery, evoking memories of emancipation and freedom, but this may only be because the addressee of the characterization (Deckard) is assumed to be a human, who may therefore find the actual *Replicant* existence unintelligible in any other terms. Indeed, Roy says as much, repeatedly emphasizing that humans and *Replicants* occupy different universes. As Roy puts it to Deckard: 'I've seen things you people wouldn't believe. . . .' Yet the intonation here is more of despair than of accusation; evoking the injustice of being unable to put into phrases what *ought* to be presented in the available idioms (Lyotard, 1988; Lyotard and Thébaud, 1984). And even whilst this remark smacks of despair, it echoes Roy's earlier encounter with the genetic creator of his eyes at the Eye Works, where he jokes about precisely the same situation, remarking 'If only *you* could have seen what *I've* seen with *your* eyes.' It seems far from certain, therefore, that the renegade *Replicants* want to simulate perfectly the human condition (and so become indistinguishable from humans), or to have the same 'rights' as humans (and so become equal to humans), or even really to become human (whatever that might mean). To the contrary, we suspect that what the *Replicants* want is something that is otherwise-than-being. Specifically, they want something otherwise than being-towards-death (which would be anything but human). We shall return to this under the themes of undecidability and symbolic exchange below. Yet as we shall see, undecidability and symbolic exchange are interlaced within a specifically disadjusted temporality.

As a way into the question of temporality, consider Jameson's (1983) linking of postmodernism to consumerism. The *Replicants* are characterized by built-in obsolescence. Their being-towards-death is literally encrypted into their flesh (cf. Oncomouse™; Haraway, 1992). And it is the difference between their restricted four-year lifespan and their desire for an unrestricted longevity that motivates the narrative of *Blade Runner*: they want to be de-crypted and re-coded. In other contexts, one might say that they want to be understood. For the moment, however, we will stick with Bruno's suggestion that the restricted lifespan of the *Replicants* accords with a specifically postmodern temporality: 'The replicant affirms a new form of temporality, that of schizophrenic vertigo' (Bruno, 1987: 69).

On Jameson's account, schizophrenia is the result of a language disorder, the schizophrenic being 'condemned to live in a perpetual present' as a result of a breakdown in the normal process of accession into the symbolic order (Jameson, 1983: 119). Since language signifies by means of syntagmatic chains of signifiers, subjective experience of temporality – specifically its linear continuity, running from the past, through the present, to the future – is immanent to signification. Hence, the failure of the schizophrenic to accede properly into language accounts for the disadjusted experience of time as a 'perpetual present'. But this discontinuous experience of time is simultaneously endowed with a shimmering intensity unbeknownst to normal subjectivity, where the present 'is always part of some larger set of projects which force us selectively to focus our perceptions' (*ibid.*).[10] Some of this arguably accords with *Blade Runner*'s representation of *Replicant* subjectivity: as Tyrell remarks to Roy, 'a light that burns half as long burns twice as bright. And you have burned so very, very brightly.' On this reading, the *Replicants*' desire to overturn their artificially restricted lifespan amounts to a desire for normal subjectification. Pris, however, appears to insist on this latter in advance: 'I think, Sebastian, therefore I am' is her response to J. F. Sebastian's indelicate request to 'show' him what they, as *Replicants*, can do – and, as Lacan (1977) has argued, the *Cogito* is the exemplary, rationalized manifestation of the *méconnaissance* by which the subject misidentifies itself as fully coherent and self-present. The *Replicants*' apparent striving after a normal, human subjectivity would thus amount to a desire to accede into the symbolic order, necessitating an 'Oedipal journey' (Bruno, 1987: 71).

Developing this line of thought, Silverman (1991) has detailed *Blade Runner*'s psychoanalytic resonances. Rachel, for example, is an experimental *Replicant* that has been programmed with artificial memories in an effort to prevent her own self-recognition as a *Replicant*: 'If we give them a past, we create a cushion or pillow for their emotions, and consequently we can control them better,' remarks Tyrell. When the blade runner, Deckard, recalls for Rachel 'her' memories of playing 'doctor' with her brother ('He showed you his, and when it got to be your turn you chickened and ran'), and of the spider that lived outside her window (whose egg hatched, yielding 'a hundred baby spiders', who proceeded to devour their mother), the implication is clear: 'Those aren't your memories, they're someone else's; Tyrell's niece,' states Deckard. Rachel musters a photograph in her defence, which she imputes to represent herself as a child with her mother, thus providing positive proof of her authentically human identity.[11] As Silverman (1991: 120) notes, these elements 'provide Rachel with an entire . . . Oedipal history', through which she has been cybernetically constituted as a subject – following precisely the same trajectory as the human child engendered as a female subject.[12] The way in which Rachel repeatedly submits to Deckard reinforces this. Replaying the Oedipal drama, Rachel re-assumes the definition of Woman (*La Femme*) in accordance with the phallocentric ordering of the symbolic.

In contradistinction to Rachel, however, the *Replicant* leader Roy – who exhibits what according to Bruno (1987) is the *Replicant*'s form of schizophrenic subjectivity – clearly fails to submit to the Name-of-the-Father, in that he blinds, then murders, Tyrell. As Silverman (1991: 121) notes, 'Not only does he literally murder the figure who produced him, but he kisses him passionately first, in an astonishing condensation of both the negative or homosexual, and positive or heterosexual versions of the male Oedipus complex.'[13] According to Silverman (*ibid.*), it is 'his refusal to relinquish one [version of the male Oedipus complex] on behalf of the other' that results in Roy's inability to undergo a process of Oedipalization, which culminates in patricide and the subsequent acceptance of terminal breakdown. Whilst Silverman (*ibid.*) sees Roy's patricidal actions as the 'unpredictable consequences' of 'Tyrell's dream of controlling the replicants through their implanted memories' – revealing that Tyrell is 'profoundly misbegotten' – this neglects the fact that it is Rachel, and not the renegade *Replicants*, who has the particular status of an 'experimental' *Replicant* with artificial memories. And whilst Leon has a collection of photographs – which Deckard does equate with Rachel's, judging them to be 'just as phony' – Leon's do not appear to be Oedipalizing 'family snapshots' in the manner of Rachel's. One of Leon's photographs, for instance, is a shot of a seemingly empty hotel room – which, when Deckard subsequently analyses it using an Esper machine (in an obvious reference to Antonioni's *Blow-Up*: see Easthope, this volume), reveals the shimmering image of Zhora, reflected (ironically) in a mirror. Moreover, unlike Rachel, the delusion that they are authentically 'human' is absent in the case of the rebellious *Replicants*. Nor are we told that Rachel has (knowledge of) a limited lifespan.[14] The renegade *Replicants* know what they are. But as for Rachel . . .

DECKARD   SHE DOESN'T KNOW?
TYRELL   SHE'S BEGINNING TO SUSPECT, I THINK.
DECKARD   SUSPECT! HOW CAN IT NOT KNOW WHAT IT IS?
TYRELL   COMMERCE IS OUR GOAL HERE AT TYRELL; 'MORE HUMAN THAN HUMAN' IS OUR MOTTO.

(*Blade Runner*)

For Bruno (1987), however, the sense of schizophrenia is not restricted merely to its personification in the form of the *Replicants*. Rather, the presence of the *Replicants* serves to affirm 'the fiction of the real' (*ibid.*: 67) throughout the diegesis of the film. The category *Replicant*, one might say, amounts to 'a deterrence machine set up in order to rejuvenate the fiction of the real in the opposite camp' (Baudrillard, 1994: 13). For Baudrillard (1988: 27), 'The schizophrenic is not, as is generally claimed, characterized by a loss of touch with reality, but by the absolute proximity to and total instantaneousness with things.' He is 'Stripped of a stage and . . . cannot produce the limits of his very being, he can no longer produce himself as a mirror' (*ibid.*). Hence, the stable *méconnaissance* of subjectivity and representation characteristic of modernity are short-circuited by the postmodern screening of an open and deterritorializing schizoid subjectivity, which is always already becoming-other through 'a being-multiple, instead of a being-one, a being-whole or being as subject' (Deleuze and Parnet, 1987: viii; Doel, 1995). And yet, the necessity of the 'retirement' of the renegade *Replicants*, and the obsession with enforcing the distinction between them and humans in the film, speaks more of paranoia (the pathology of organisation) than of schizophrenia. So, whilst Bruno suggests, with respect to Baudrillard's (1994: 2) account of the simulacrum as 'an operational double, a programmatic, metastable, perfect descriptive machine that offers all the signs of the real and short-circuits all its vicissitudes', that 'it would be difficult to find a better definition of the nature and function of the replicants and their capacity of simulation in the narrative function of *Blade Runner*' (Bruno, 1987: 68), this is to miss the obsession with, precisely, the *difference* between *Replicant* and human within the film. Indeed, as we shall go on to argue, much of the film is concerned with establishing, testing and verifying the difference between real, authentic human life on the one hand and the *Replicant* simulation of human life on the other: organic presence (to live, experience, remember and die) *versus* machinic reproduction (to function, operate, encode and break down).

Accordingly, whilst *Blade Runner*'s component parts might appear to mirror certain aspects of our own allegedly postmodern epoch – where space-time has become qualitatively different through the fissiparous process of de-differentiation, which serves to bring different elements into almost absolute proximity, thereby threatening to short-circuit and earth their polarity and charge – the film's diegesis is far from breaking with the desire to assign each element an appropriate identity or location within the Order of Things. To the contrary, both the space-time of *Blade Runner*'s diegesis and its cybernetic mimicry of subjectification remain conventional, banal, and predictable inasmuch as they are essentially combinatorial and kaleidoscopic. As with 'compression', the margins of difference may have been diminished, but everything remains frozen in the categories of the Same. For all the supposedly heterotopic juxtaposition and superimposition of diverse elements in social space, the resultant amalgam offers no resistance to comprehension and analytic dissection: bodies remain divisible by gender, ethnicity, class and brain structure; the cityscape remains divisible into discrete zones (the Eye Works factory; the Tyrell Corporation Building; Apartment blocks; China Town); temporality remains an irreversible, universal and linear flow (where, subjectively, its end – terminal breakdown or death – has come to determine the meaning of life); and images and implants remain inauthentic in relation to experience and memory. In short, the account of the postmodern supposedly reflected in *Blade Runner* does nothing less than domesticate it, in line with a thoroughly modernist logic. This is not a world of undecidability, clandestinity, and becoming, but of massive over-determination. *Replicant* and human are not categories which emerge from a cognizing of the world; they are order-words which must be re-cognized and obeyed.

To escape such a logic, the '*Post modern* would have to be understood according to the paradox of the future (*post*) anterior (*modo*)' (Lyotard, 1984: 81): it *will have been*. And whilst this disadjusted temporality may seem to resonate with *Blade Runner*'s contraction of time, space-time would here embody a truly fractal quality; an alchemical and lycanthropic becoming-other – rather than a rigid grid for the location and unfolding of immutable beings – which differs and defers the presences and identities upon which all architectures of periodization rest (Doel, 1992). Hereinafter, 'There is no more system of reference to tell us what happened to the geography of things' (Baudrillard, 1987: 126). For us, however, *Blade Runner*'s combination of *film noir* and science fiction engenders nothing more than a context ripe for nostalgic extrapolation: a perfect backdrop for a homely yet heroic stroll through a trying rather than a malicious world. The fusing of detection and invention ultimately ensures that nothing will escape recognition and comprehension; everything comes back to a dissimulation of truth behind a veil of counterfeit images and false mirrors. *Blade Runner* will have been continuously and eternally returned to the safety, order, stability, and coherence of the modern.

EX-TERMINATING THE SIGNS OF LIFE

'In history books he's the kind of cop who used to call black men niggers.' So remarks Deckard – in one of the voice-overs subsequently removed from the Director's Cut – commenting on the reference by Bryant, his boss, to *Replicants* as 'skin-jobs'. But, despite the film's potential for a 'racialized' development of the difference between human and *Replicant* – evident in other sci-fi films, such as the *Alien* trilogy – it is arguably in relation to sexual difference that *Blade Runner*'s narrative operates.[15] According to Penley (1991: 72), science fiction film frequently displaces questions of sexual difference 'on to the more remarkable difference between the human and the other'. And in the case of *Blade Runner*, the *Replicant* clearly adopts the position of Woman – repeating Woman's 'originary displacement', conceived as the capacity for masquerade (Riviere, 1929). It was Nietzsche (1974), arguing within the historical understanding of his time, who suggested that Woman, supposedly incapable of orgasm, derives pleasure soley from the *impersonation* of sexual ecstasy. As Spivak (1983: 170) puts it: 'At the time of the greatest self-possession-cum-ecstasy, the woman is self-possessed enough to organize a self-(re)presentation without an actual presence (of sexual pleasure) to represent. This is an originary displacement.'[16] Within *Blade Runner*, the difference played out between human and *Replicant* parallels exactly this structure of displacement. The *Replicants* are ultimately distinguished by their self-possessed capacity to 'fake it' – to simulate (is) their own being. Moreover, as Deckard's manliness is repeatedly questioned throughout the course of the film, this amounts to the displaced questioning of his human status. This occurs principally through the character of the police officer, Gaff, who at three different points in the film secretes tiny symbolic models within Deckard's immediate vicinity: first, a minute origami chicken, when Deckard is reluctant to take on the assignment of 'retiring' the renegade *Replicants*; second, a phallic match-stick man, when Deckard searches Leon's hotel room; and, finally, a tiny origami unicorn. But Roy, too, appears to cast such aspersions, saying at one point in his final encounter with a gun-toting Deckard, 'You'd better get it up or I'm going to have to kill you.' After this final encounter between Deckard and the *Replicants*, which leaves Pris and Roy dead but Deckard alive, Gaff congratulates Deckard, saying 'You've done a man's job, sir'[17] – and then immediately redoubles the ambiguity of Deckard's identity by commenting, with respect to Rachel: 'It's too bad she won't live. But then again, who does?'[18] The supposedly determinate distinction between human and *Replicant*, therefore, is above all produced by a categorization that perpetually breaks down. And, as we shall see, its significance is established in the film through the recurrent motif of the Eye.

The *Replicants* produced by the Tyrell Corporation are, first and foremost, *products*. Moreover, they are mass produced products, which roll off the assembly line according to precise specifications and by way of a complex network of subcon-tractors, evidenced, for example, in the small 'Eye Works' factory tucked away in the backstreets of Chinatown. As a product, then, the industrially produced *Replicant* lends itself to all of those nostalgic laments for a by-gone age of authen-tically crafted *works* – one-off, spontaneous, and context-bound productions, whose unique quality and originality can only be counterfeited. Needless to say, such a counterfeit will always be inferior to, and parasitic upon, its original. From such a perspective, the industrially produced product is invariably judged to be a poor imitation of, and somehow less worthy than, the original work which it comes to supplement, and increasingly to displace. According to this script, then, the *Replicant* appears as the ultimate moment in the colonization of the lifeworld by counterfeited signs of life (Kearney, 1988; Lefebvre, 1991). The *Replicant* stands as a post-humanist figure for an inhuman(e) world – but only on condition that this figure remains flawed in relation to the wholesomeness of real (human[e]) life.

In this way, the industrially produced *Replicant* also lends itself to all of those critiques of everyday life under capitalism as one of alienation, reification, and fetishism, so beloved by humanists and Marxists alike. The *Replicant* is not a product that is unintelligible to us, despite its 'animation'. To the contrary, it is all too familiar – a product *like any other*. In fact, the *Replicant* is a commodity *par excellence* – since, like all commodified objects, *Replicants* have business, commerce, and intercourse amongst themselves. They take on a life of their own, and converse with themselves in the market of equivalences. To that extent, the *Replicant* is a perfect encapsulation of the vampiric form of capital accumulation – the ossified out-turn of past human labour feeding off the present toil of living human labour. *Replicants*, like capitalism itself, are nothing but so much dead labour simulating life. All of this fits neatly into the common-or-garden desires of so-called modernists: they want their objects, products, commodities, and machines to be efficient, exchangeable, durable, and on occasion intelligent – but they also want them to be deprived of their own will, desire, sexuality, and destiny (Baudrillard, 1995). They want their objects to function *as objects*, and never as subjects. For the modernist, the only good object is a dead object: cold, passionless, and calculative – a machine for (the) living; rather than a truly living machine (Latour, 1992).

From the modernist perspective – whether nostalgic, humanist, or Marxist – the interest of *Blade Runner* derives largely from the *Replicants'* apparent transgression of the pact between humans and objects. The *Replicants* do not simply express the fetishism of commodities through the adoption of so many social relations amongst themselves, they also aspire to become fully fledged subjects, thus erasing the qualitative distinction between real (human[e]) life and the counterfeited signs of life. The transgression which the *Replicants* threaten is therefore not between good and bad copies of an original; whether in terms of the first, natural form of simulation or the second, industrial form of simulation; but between copying as such and the third, simulacral form of simulation (Baudrillard, 1993b, 1994). In aspiring to become pure simulacra, the *Replicants* are endeavouring to sever themselves from all reference to an original, and thus to attaining a specificity all of their own. For a simulacrum cannot be re-turned to *its* original, in so far as it is that for which there is neither original nor equivalent. A simulacrum has only itself. No longer a copy of something other than itself, a simulacrum is that which is always already reproduced. Indeed, it is this possible, retroactive occlusion of an original, human(e) life which serves as the all-pervasive sense of ambient fear that saturates *Blade Runner*. Lurking in the background is the fear that 'we' are all *Replicants* now – that we will never have been 'human', let alone 'modern' (Latour, 1993).

With typically modernist bravado, *Blade Runner* seems to want to assure us that *Replicants* are not simulacra but copies; and that, by the same token, 'we' remain something other than mere signs of life: human(e) subjects; living beings. And just to reassure us of the difference between humans and *Replicants*, active subjects and (in)animate objects, *Blade Runner* brings centre stage a certain test that can detect and verify the difference. This test – the Voigt–Kampff test – seems to be the last line of defence between the logic of originals and copies on the one hand, and the undecidability of simulacra on the other. Yet as we shall see, this line of defence, like the distinction itself, is an illusion.

The modernist reading of *Blade Runner* rests on the assumption that *Replicants* are near perfect *copies* of human beings, and that in many respects – such as strength, constitution, and agility – they surpass human capabilities. 'We're no computers,' Roy tells J. F. Sebastian, 'we're physical.' As a simulation, the *Replicant* is in a sense hyperreal – 'More human than human,' as Tyrell puts it in the film. And yet, no matter how perfectly the *Replicant* copies the human, there will always remain a fundamental distinction between the two. *Replicants* and humans are inalienable and immutable forms; one cannot *really* become the other, but can only enter into relations of analogy – it is *as if* the product *were really* human, but we know that it is not. Given that the renegade *Replicants* have superhuman powers – Roy smashes through walls and withstands being battered by iron bars; Leon's hand is immersed into liquid nitrogen without ill effect; and Pris' hand bears being plunged into boiling water – it may seem somewhat implausible that the authorities should resort to the time-consuming and essentially probabilistic Voigt–Kampff test for discerning the difference between *Replicants* and humans. Wouldn't a more straightforward physical test be more appropriate – such as exposure of the skin to temperature extremes, or detection of the product's serial number genetically etched on to its component parts? Indeed, the Philip K. Dick novel (1972) on which *Blade Runner* is based refers to a bone marrow test and to the 'Boneli Reflex Arc Test', which rests on the fact that response to an electrical stimulus in the upper ganglia of the spinal column is fractionally slower in *Replicants* than in humans. However, none of these tests would suffice in the long run because the manufacturers are working according to the premise that these more obvious signs will be systematically erased from their products (Alliez and Feher, 1989). Indeed, the Dick novel makes it absolutely explicit that the Rosen Association – the counterpart of the film's Tyrell Corporation – is seeking to produce products that will be undetectable and indiscernible. Once produced, they will no longer 'be' *Replicants* at all. Rachel, for example, as an experimental Nexus 6 *Replicant*, is hyperreal not through superlative powers, but through being all too human. Repeatedly, the film dwells on 'her' human flaws, fallibilities, and fears (and, ironically, she seems to attain such a 'human' status in Deckard's eyes by killing the *Replicant* Leon, who attacks Deckard immediately after he has 'retired' Leon's partner, Zhora).

Yet Deckard, too, is beginning to suspect that he may not be who he is – or so we are led to think. And some of the more explicit signs of this creeping doubt paralysis were expunged from the version of the film that went into general release, only to be reinstated in the Director's Cut. Clearly, both the audiences at the film's disastrous test previews and the Ladd Company (who distributed the film through Warner Brothers) were unnerved not only by the implication of Deckard's likely *Replicancy*, but also by the refusal of the film to make a final arbitration (Instrell, 1992; Salisbury, 1992). Like so many moderns, they were no doubt under the illusion that order, stability, and constancy are the rule, rather than the (forced) exception; and that undecidability *should* give way to a final solution (Clarke *et al.*, 1996; Doel, 1994, 1996).

The key sequence to be reinstated in the Director's Cut was Deckard's dream featuring a unicorn. As noted above, the last of the models Gaff places for Deckard to come across is a tiny origami unicorn, which Deckard finds as he and Rachel flee his apartment. His memory of Gaff's last words – 'It's too bad she won't live. But then again, who does?' – are replayed in voice-over, and the implication is that Deckard's dreams, like Rachel's memories, are not private and personal but issue from some kind of implant – such as those manufactured by the Tyrell Corporation, or the collective unconscious, for example. Gaff has seemingly had access to Deckard's dreams – or has dreamt them himself – implying that Deckard, too, is a *Replicant*.[19] And it is perhaps no coincidence that this origami unicorn is folded from a relatively lustreless piece of tinfoil, whose multiple folds do not so much focus the light into a coherent whole, as separate it into an infinitely disadjusted, spectral presence. It is as if this final sign left by the agents of the dominant, molar order were alluding to the aporetic status of the *tain* of the mirror, which is the condition of possibility of visibility, reflection and recognition, and yet remains invisible and unrepresentable within the mirror-play of identity itself (Gasché, 1986; Kearney, 1988). And in so far as the tinfoil unicorn alludes not to a secure *ontology* of unchanging forms, but to the undecidable *hauntology* of real becomings, the agents of the molar order admit not only the contingency of their own Will to Power, but also that responsibility can only take place by way of the undecidable – without which, responsibility would simply be determination, deontology, and obeyance. In their undecidability, which suspends a final solution, Deckard and Rachel respond responsibly to each other (Derrida, 1988). But we have moved too fast, since we have not yet explained why Deckard and Rachel will have been undecidable rather than determinable as either this or that; as either human or *Replicant*.

Let us return to the assumption that the Voigt–Kampff test remains the most plausible basis for discerning the difference between Humans (the Real) and

Replicants (the Hyperreal). This test relies on a table-top apparatus, which records the continuous variation of 'capillary dilation in the facial area' and 'fluctuations of tension within the eye muscle' as indexical signs of the body's involuntary response to a series of hypothetical questions that are designed to elicit an emotional response from the subject. The questions conjure scenes that are supposedly 'morally shocking' to human beings: cruelty to animals; infidelity; lesbianism; and so on. By cross-referencing these data, the operator should be able to determine whether or not the subject is a *Replicant*, since *Replicants* are deemed to lack an adequate empathetic capacity. The key word here is *should*, since some *Replicants* have already acquired enhanced empathetic capabilities, whilst some humans – most notably schizophrenics and sociopaths – lack a 'normal' dose of empathy. Empathy is not a genetically given trait; it is a product of effective socialization and normalization. Hence the significance of the fact that Rachel's stock of Oedipalizing memories were *implanted*, post production – 'it' was cybernetically socialized into 'she', and so quite literally en-gendered. (But then again, who isn't?) Likewise, in so far as the Voigt–Kampff test is spoken, it relies on an ideal speech situation of precise and transparent communication. Both Leon and Rachel, however, attempt to subvert the testing procedure through the deployment of copious quantities of 'semantic fog'. In effect, they attack the medium itself, rather than merely engaging with the message. Rather than offer an answer to a set question, they respond in such a way as to render the question itself unintelligible.

Nevertheless, the Voigt–Kampff test seems to be able to distinguish a Nexus 6 *Replicant* from a human. Whilst a conventional Nexus 6 *Replicant* such as Leon is seen to require only twenty to thirty cross-referenced questions to disclose itself, an experimental one such as Rachel requires over a hundred. Thus, after ten questions both Leon and Rachel were confirmed neither human nor *Replicant* – they were both undecidable. After thirty questions, Leon had been identified as a *Replicant*, but Rachel was still undecidable. After over one hundred questions, Rachel was confirmed as a *Replicant*. But what of another subject? After how many questions should they be confirmed *human* – two hundred? One thousand? Ten thousand? Paradoxically, the Voigt–Kampff test can *never* confirm a human being as human; it can only produce *Replicants* (the falsified) and Undecidables (the not-yet-falsified) at an arbitrarily set level of resolution.

It should be evident that at no single point is the status of a body disclosed, since the test relies on cross-referencing. Rather, disclosure is probabilistic – and this probability varies continuously in relation to the accumulating data. As we have seen, any decision requires the arresting of the test at some arbitrarily defined point, in so far as subsequent data may always suggest otherwise. Further questioning may evoke shock and empathy – or not. Still further questioning may alter the probability distribution – or not. And so it goes, on and on interminably. This is why the test result is quite literally a resolution of undecidability; it re-releases and unfastens it again. The arbitrarily foreclosed determination opens in and of itself on to indeterminacy and undecidability. It is a mistake to believe, therefore, that the Voigt–Kampff test leads a body to one of two terminals: either *Replicant* or Undecidable; and still less to either *Replicant* or Human. What it actually does is *suspend determination in the realm of the undecidable* because the apparently falsified may become *un*falsified at a higher resolution, and vice versa. Paradoxically, then, rather than returning a body to its essential terminus, the Voigt–Kampff test literally ex-terminates the difference. Since the test is probabilistic, it only ever produces undecidability. Such a state of radical exception is synonymous with terror, a state which suffuses the quotidian with ambient fear (Clarke and Doel, 1994; Doel and Clarke, forthcoming). Hereinafter, one will never know if a body is *really* a human or a *Replicant* – only whether or not it expresses signs of *Replicancy* or humanity at various levels of resolution.

Here we arrive at the central paradox of the film. Whilst the leading protagonists agonize over the status of (their) bodies, the dominant order remains ambivalent. As we have seen, the Voigt–Kampff apparatus does not so much detect difference as erase it – but it erases it in a very particular way. By retroactively negating the essence of things through a wave of undecidability, the dominant order produces its own product – an amorphous surface upon which it can impose order. In this sense, the Voigt–Kampff test is a perfect encapsulation of what Deleuze and Guattari (1988) call the 'Terrible Ray Telescope', which is not used to see with, but to cut with. (In this sense, the Will to Truth is nothing but the Will to Power.) The test apparatus acts on the flesh, cutting out identifiable figures that can be accepted or rejected at various – arbitrarily set – resolutions. It is no small coincidence that the Voigt–Kampff apparatus works by piercing the subject's eye with a beam of light. Moreover, as if to underscore the ambivalence of the established order towards the so called 'reality' of the referent, it is worth noting that the monitor attached to the Voigt–Kampff apparatus always renders the subject's eye as green, even when in actuality it is not (Leon's eyes are blue, Rachel's are brown; Silverman, 1991).

The Voigt–Kampff test, with its focus on the Eye – and the I – is, therefore, absolutely pivotal to the world of *Blade Runner*. As it recurs throughout the film, the test amounts to the moment where the forces of the molar order attempt to normalize and stabilize a single instance of the continuous variation of a molecular flow – a seizure of power that overcodes and normalizes in its attempt to encode everything in its own image. It is the imposition of a statistical norm or majoritarian standard around which deviations and deviances will be distributed. As Lyotard reminds us, 'majority does not mean great number but great fear' (Lyotard and Thébaud, 1984: 93; Deleuze and Guattari, 1988). And so, as we have already noted, the Voigt–Kampff test is precisely *indifferent* to the actual status of any body's identity. There is no fixed, authentic human identity; no fixed (in)authentic *Replicant* identity. Both are held in a suspended state of indeterminacy and undecidability; both only ever will have been. Thus, the Eye/I becomes the terrain upon which the molar forces of order do battle with the molecular forces of becoming, dissemination, and undecidability. The stability of identity, which the molar order persistently attempts to impose, always already fails. Hence, the film is entirely consumed by the tension between a yearning for clearly defined and immutable identities – figured by the eternal recurrence of an interminable blink, which ceaselessly strives for, yet ultimately suspends, the self-secured closure of the Eye/I around itself – and the fact that there are no inalienable forms or fixed partitions – as figured by the explosion of becomings along the dimension of the Eye/I (cf. Bataille, 1982; Barthes, 1982).

Blade Runner thus proceeds through a screening of the Eye/I, which is always already other than what it will have been. Indeed, from its very opening shot, the film is saturated by this metamorphosis. Not only does the centrally important Voigt–Kampff test focus on the eye; the apparatus itself has eyes. In addition: vehicle headlights and street lights constantly evoke eyes; the eyes of the *Replicants* glow menacingly or mysteriously (Deckard's eyes at one point glow in this way); in the Eye Works factory, Leon places genetically engineered eyes on the shoulders of a terrified worker, whose thermal suit resembles nothing so much as an eye, its dangling optic nerve – the suit's electrical cables – severed by the *Replicants*; when Deckard questions Zhora about her treatment as an entertainer he refers to tiny peep-holes drilled into the dressing room walls; eyes transmogrify into eggs in the scene where Pris – who has painted out her eyes – plunges her hand into the pot of boiling water on Sebastian's stove; Roy jokes with Sebastian by holding glass eyes in front of his own; the powerful, varifocal lenses of Tyrell's glasses magnify his eyes before Roy blinds him – recalling Leon's attempt to put out Deckard's eyes after his 'retirement' of Zhora; in one of Leon's photographs, Zhora is reflected in a fish-eye mirror; and so on. Hence, from the opening reflections of the cityscape in the false mirror of an unblinking Eye, to the closing sequence of the Director's Cut in which Deckard and Rachel are enveloped by the contracting aperture of an elevator shaft, the Eye proliferates across the screen. To that extent there is a homology between the undecidable, disseminating Eye and the undecidable, disseminating I. And this twofold uprooting of the centred subject brings us to perhaps the most puzzling dimension of the film: Death. But as with Eyes and Is, this is not really death of a mortal kind; nor is it the death of an individual. As we shall see, death is always already both symbolic and collective. And it is this which enables the apparent apoliticism of *Blade Runner* to offer a glimmer of emancipatory promise.

'TO LIVE IN FEAR': THE REVERSIBILITY OF LIFE IN DEATH

There is a strange dialectic played out by Roy and Deckard towards the end of the film, which hangs on the reversibility of life in death – and the way in which capital has sought to render this relation irreversible. It concerns the status of the *Replicants'* lives as slave labour in the Off-world colonies. In *Symbolic Exchange and Death* (1993b), Baudrillard opposes labour as slow death not to 'the "fulfilment of life", which is the idealist view' (*ibid.*: 39), but to *violent* death: 'Labour is opposed as deferred death to the immediate death of sacrifice' (*ibid.*). This he derives from the genealogy of the slave:

> First, the prisoner of war is simply put to death (one does him an honour in this way). Then he is '*spared*' [*épargné*] and *conserved* [*conservé*] (= *servus*),

under the category of the spoils of war and a prestige good: he becomes a slave and passes into sumptuary domesticity. It is only later that he passes into servile labour. However, he is [not yet] a 'labourer', since labour only appears in the phase of the serf or the *emancipated* slave, finally relieved of the mortgage of being put to death. Why is he freed? Precisely in order to work.

(Baudrillard, 1993b: 39)

Hence, we can distinguish between two regimes of death: the *economic* (the attempt to render life an irreversible form) and the *sacrificial* (founded on the reversibility of life in death). 'We live irreversibly in the first of these,' notes Baudrillard, 'which has inexorably taken root in the *différence* of death' (*ibid.*). Hence, labour amounts to being judged worthy *only* of life; capital primarily inflicts upon the labourer the refusal of death. Accordingly, power is defined not by the taking but by the giving of life; not by the giving of death but by its taking – its refusal or deferral (Derrida, 1992). This is pointed up far more clearly in the case of the *Replicants* than in human labour, inasmuch as the *Replicants*' function as labourers is their sole reason for being produced; their very existence is a poisonous gift from capital in the shape of Tyrell. Thus, as Baudrillard notes,

Contrary to all appearances and experiences (capital buys its labour power from the worker and extorts surplus value), capital gives labour to the worker. . . . In German this is *Arbeitgeber*: the entrepreneur is a 'provider of labour'; and *Arbeitnehmer*: it is the capitalist who gives, who has the initiative of the gift, which secures him, as in every social order, a pre-eminence and a power far beyond the economic.

(Baudrillard, 1993b: 41)

The symbolic relation of the gift is thus the locus of terror: as Roy says to Deckard – echoing but completing Leon's earlier words – 'Quite an experience to live in fear, isn't it? That's what it is to be a slave.' Indeed, Roy sardonically remarks, as he enters Tyrell's room in a doomed attempt to request of Tyrell an extention to his life, 'It's not easy to meet your maker.' This strange return of the prodigal son almost literally brings home the message that the whole of a *Replicant*'s life is lived on 'given' or 'borrowed' time. The rebel *Replicants* have come to Earth to extend their credit, but in doing so would merely increase their debt to capital, to Tyrell.

The symbolic act that Roy goes on to commit – the murder of the Father – thus goes beyond the Lacanian symbolic. And it does so in precisely the direction of Baudrillard's concerns with symbolic exchange: '*The refusal of labour, in its radical form, is the refusal of . . . symbolic domination* and the humility of being bestowed upon' (*ibid.*: 41). In murdering Tyrell, Roy performs the symbolic act of liberating himself from 'The gift and the taking of labour [which] function directly as the code of the dominant social relation, as the code of discrimination' (*ibid.*) – thereby gaining a kind of subjectivity, defined in relation to his own death. It is only following this act that the encounter between he and Deckard can take on the form it does. In effect, Roy's subjectivity is that of something *more* than just a (hu)man.

In the sequence in which Roy chases Deckard through the deserted Bradbury Building, Roy undoubtedly enacts a cruel revenge on those social forces which enslaved him. Roy himself enslaves Deckard – the agent *par excellence* of the molar order – but this is complicated by Deckard's role as, in effect, a sacrificial agent. Roy is no longer struggling to catch up with Deckard, struggling to take his life; he surpassed him long ago. Rather, Roy *gives* Deckard time – time to live; and time to die. It is Deckard who is now living on borrowed or given time, as Roy exerts the power he has obtained from his symbolic challenge to the dominant order. Roy removes the immediacy of Deckard's death, and holds it over him. Here, for the first time in the film, the sacrificial regime of death in which Deckard, as a blade runner, is implicated, surfaces explicitly. And its primordial status is alluded to in the primal screams issued by Roy and, to a lesser extent, Deckard in this scene.

By giving Deckard life, therefore, Roy does not simply adopt the position of the Father in the (Lacanian) symbolic order. Rather, he affirms a symbolic register that cannot be 'circumscribed . . . within an individualised unconscious, thus reducing it, under the Law of the Father, to the obsessional fear of castration and the Signifier' (*ibid.*: 1). Indeed, Roy's rebellious characteristics cannot be formulated in accordance with an Oedipal process in so far as he is *produced* as a *Replicant*, rather than reproduced as a human. The fact that Roy is *beyond* the human is, moreover, reinforced by the religious symbolism that is brought into effect in this scene: at one point, Roy drives a nail into His hand, ostensibly to bring about sufficient pain to ward off His final seizure and terminal breakdown. But in so doing He assumes the position not of the Father but of the Son: the symbolism of the crucifixion, of the sacrificial lamb. Accordingly, He takes His position as a subject, but a subject who has demanded the honour of sacrifice rather than subjection to the deferred death that is afforded to the life of a slave. This symbolism is, of course, highly ironic in so far as Christ was the ultimate obedient Son.

When Roy finally brings his teasing of Deckard to an end – a coaxing which forces Deckard to act like a 'real man' – He literally holds Deckard's life and death in His hands. As Deckard barely hangs on to the facia at the top of the building, to which he has attempted a death-defying leap, one might expect Roy to let him plummet to his death. One might expect Roy to secure His revenge by exchanging a life for a life, His life for Deckard's death, His immanent slow death for Deckard's temporarily suspended violent death (Plate 7.4). But Roy catches him as he falls – just as the symbolic order catches the subject – again reinforcing the relation by which Roy gives Deckard his life (takes his death). Thus, and in contradistinction to Harvey's (1989) claim that there are no alliances amongst supposed humans and *Replicants*, Roy emancipates Deckard. If Roy liberated Himself by adopting the position of the Father in enslaving Deckard, Roy now escapes the Law-of-the-

7.4   Beyond the human: Roy nears His end
© Warner Bros. Courtesy of BFI Stills, Posters and Designs.

Father by taking back the life which He had given Deckard, but in a way that also gives back the death which was taken from Deckard. Henceforth, the symbolic debt which both Roy and Deckard carry for living on borrowed time is cancelled. In the closing moments of the rooftop scene, upon which Ridley Scott plays out his Death Drive, Roy *lets himself die* as something more than a (hu)man; He gives Himself entitlement to give His own life, and to take His own death. All of this is crystallized in the *longue durée* of Deckard's blink at the film's moment of greatest intensity: the breakdown of Roy, head down, face flooded with rain and tears, and eyes sealed against the torrent of fleeing memories. And finally, here, the logic of the religious symbolism is finally consummated, as Roy's death coincides with the release of a white dove from His grip, completing the Trinity as His Spirit ascends towards the darkened heavens. He has finally overcome His *Replicant* status as a being-towards-(a slow)-death.

Paradoxically, then, there are not two endings to *Blade Runner* (1982; 1992): one of Rachel and Deckard descending into the abyss of the dead-eye of power; the other of their fleeing into those spaces that subsist beyond the reach of Los Angeles 2019. Both endings amount to the same: a deep ambivalence with respect to the null hypothesis of what it is to be(come) human; and a deep ambivalence towards its others (*Replicants*, schizophrenics, 'little people'. . .). In the lift-shaft descent into the nether regions of the 'not-yet-falsified' of 'borrowed time', Deckard and Rachel are swallowed and consumed by the dominant order. In the meandering drive through the ex-urban countryside, they are (temporarily) expelled and exiled from the dominant order – although there is nothing in the film to foreclose the possibility that this 'escape' is itself a staged experiment by the Tyrell Corporation, or that the film in its *entirety* is an implanted memory. Whether swallowed, expelled, on the run, or experimented upon, both Deckard and Rachel are given (borrowed) time, and are thereby obligated to live to the full the life that is credited to them on account of their enslavement to a slow death. 'Only the counter-gift, the reversibility of symbolic exchange, abolishes power' writes Baudrillard (1993b: 49). Only Roy manages to achieve the symbolic violence of ex-termination through His acceptance of death: 'Perhaps death and death alone, the reversibility of death, belongs to a higher order than the code' (*ibid*.: 4). There is more than a measure of irony, therefore, in the way in which virtually all existing readings of the film have missed the permeation of the world of *Blade Runner* by death, ex-termination, and symbolic exchange – and have merely contented themselves with the documentation of the sightings of the ghosts of the symbolic, around which all modern critical theories have continued to gravitate.

ACKNOWLEDGEMENTS

We would like to thank Steve Daniels and Rob Lapsley for comments on a previous version of this chapter, and Martin Gittins and John Gold for bringing particular articles on the film to our attention.

NOTES

1 Peter Wollen (1991), for instance, has cited *Blade Runner* as the only canonical film of the 1980s.

2 This situation is characterized by both an intensity and a tension – a pervasive sense of ambient fear – that saturates the city streets. Even the language of the street has evolved into a 'mish-mash' of Japanese, German, Spanish and English referred to as 'cityspeak'.

3 According to Salisbury (1992: 94), before the title *Blade Runner* was finalized, Ridley Scott 'was . . . keen on the title *Gotham City* but Batman creator Bob Kane refused permission to use it.'

4 For his part, William Gibson is obviously amused by, rather than incredulous towards, dystopic visions of Los Angeles: 'Town planners of LA have six scenarios of the way that the city could turn out and one of the worst ones is the "Blade Runner Scenario", which I think is great' (Gibson, 1993: 13).

5 Žižek (1992: 217) also notes that 'Ridley Scott's films display a vision of a corrupted and decaying megalopolis', commenting that 'in *Black Rain*, he finally stumbled upon an object whose reality itself gives body to this vision: today's Tokyo [sic] – no need, there, to take refuge with dystopian visions of Los Angeles in 2080 [sic], as in *Blade Runner*.' Scott's 'exact specifications' for the *Blade Runner* set are elsewhere recorded as: 'Hong Kong on a bad day' (see Salisbury, 1992: 96).

6 *Blade Runner*'s commentators frequently list: neo-Mayan architecture and Fritz Lang's *Metropolis* (both of which are, in fact, signifiant of a hierarchically ordered society); Great Universal Studio's 'New York Street' set, used in such films as *The Big Sleep* and *The Maltese Falcon*; the Ennis–Brown House designed by Frank Lloyd Wright; Classical, Oriental, Greek, Roman and Egyptian styles; and so on.

7 We might add Scott's native Middlesborough in the North East of England to this compilation.

8 This is precisely the point made in Baudrillard's (1993b) development of Saussure's theory of the anagram.

9 As Marder (1991: 97) notes, the term 'Replicants' shows its etymological connection to 'replicas' but becomes '*replicants* thereby echoing from the Latin the present active participle *ans*'. Alliez and Feher (1989: 55) were amongst the first to identify the *Replicant* as a *simulacrum*, 'a usurper who passes himself [sic] off as the real thing or, worse, blurs the distinction between the model [or original] and the copy.' There is an evident similarity between the *Replicant* and Haraway's (1985) 'cyborg' (cybernetic organism).

10 Jameson's reading of Lacan arguably misspecifies the *disadjusted* temporality of 'normal' subjectivity, which issues from the *retroactivity* of the process of signification.

11 A number of commentators on the film have, following Barthes (1981: 76), stressed the particular status of photography in this respect: 'in Photography I can never deny that the thing has been there. There is a superimposition here: of

reality and of the past.' As a fuller explication, Barthes (*ibid*.: 77) writes: 'The name of Photography's noeme will therefore be 'that-has-been,' or again the Intractable. In Latin, this would doubtless be said: *interfuit*: what I see has been there, and yet immediately separated; it has been absolutely, irrefutably present, and yet already deferred.' None of this is to deny, of course, that the referent may be mistakenly believed to be other than what it is – as in Rachel's case.

12 As Silverman (1991: 120) notes, in this interchange in the film the mother–daughter relation, as imaged in Rachel's photograph, is ruptured by Deckard in a manner which condenses the Oedipus complex. The game of 'doctor' metonymically reproduces the female castration complex; the spider story – a fantasy about killing and ingesting the mother – the rivalry specific to the positive Oedipus complex, where Woman is figured as lacking the phallus.

13 Freud (1962: 23) writes: 'the simple Oedipus complex is by no means its commonest form, but rather represents a simplification or schematization. . . . Closer study usually discloses the more complete Oedipus complex, which is twofold, positive and negative, and is due to the bisexuality originally present in children', explaining that 'a boy has not merely an ambivalent attitude towards his father and an affectionate object-choice towards his mother, but at the same time he also behaves like a girl and displays an affectionate feminine attitude towards his father and a corresponding jealousy and hostility towards his mother.'

14 Such differences are a source of disagreement between Massey (1991) and Harvey (1989). We would note that, whilst the gender politics of the asymmetry between Rachel's and Roy's Oedipalization is notable, Harvey has picked up accurately on this aspect of the film.

15 Both Marder (1991) and Silverman (1991) discuss this issue a little, but most of their attention is given over to sexual difference.

16 This relation between the originary displacement of Woman and the *Replicants* was first suggested to us by Antony Easthope at the 1993 Leeds International Film Festival 'Screenscapes' Conference.

17 'You've done a man's job.' This could mean: Yes, you really are a man; Yes, having done a man's job, you have proved yourself a man; Yes, having done a man's job, you have proved yourself to be equal to a man; Yes, having done a man's job, you have become a man; or Yes, you, who are not a man, have done a job which should have been a man's job.

18 'She won't live.' This could mean: She will be *killed*; She will be '*retired*' by a blade runner; She will be '*retired*' at the expiry of her restricted lifespan; She will *break down* when her components fail; She functions, experiences, empathizes, etc., but this is not *living*; She will not be *allowed* to live – only to function, exist, survive, etc.; She *refuses* to live; She *will* refuse to live; She refuses to live life as it *should* be lived; or She, Rachel, will no longer be the *same woman* after this experience. 'But then again, who does?' of course transfers all of these possibilities not only on to Deckard but on to everyone else as well.

19 We are agreed, then; the Director's Cut makes it abundantly clear that Deckard really is a *Replicant*. And yet, as we shall see, things are not quite so straightforward; not least because Roy and Deckard fail to re-cognize each other until the bitter end – it is as if their beings-in-the-world were entirely different.

REFERENCES

Aglietta, M. (1979) *A Theory of Capitalist Regulation: The US Experience*, London, New Left Books.

Alliez, E. and Feher, M. (1989) 'Notes on the sophisticated city', in M. Feher and S. Kwinter (eds) *The Contemporary City: Zone 1 and 2*, New York, Zone Books, 41–55.

Barthes, R. (1981) *Camera Lucida*, New York, Hill and Wang.

Barthes, R. (1982) 'The metaphor of the eye', in G. Bataille, *The Story of the Eye*, Harmondsworth, Penguin.

Bataille, G. (1982) *The Story of the Eye*, Harmondsworth, Penguin.

Baudrillard, J. (1987) *Forget Foucault*, New York, Semiotext(e).

Baudrillard, J. (1988) *The Ecstasy of Communication* New York, Semiotext(e).

Baudrillard, J. (1993a) *The Transparency of Evil: Essays on Extreme Phenomena*, London, Verso.

Baudrillard, J. (1993b) *Symbolic Exchange and Death*, London, Sage.

Baudrillard, J. (1994) *Simulacra and Simulation*, Ann Arbor, University of Michigan Press.

Baudrillard, J. (1995) 'The virtual illusion: or the automatic writing of the world', *Theory, Culture and Society* 12: 97–107.

Bruno, G. (1987) 'Ramble city: postmodernism and *Blade Runner*' October 41: 61–74.

Clarke, D. B. and Doel, M. A. (1994) 'Transpolitical geography' *Geoforum* 25(4): 505–24.

Clarke, D. B., Doel, M. A. and McDonough, F. X. (1996) 'Holocaust topologies: singularity, politics, space', *Political Geography* 15(6/7): 457–89.

Culler, J. (1988) *Framing the Sign: Criticism and its Institutions*, Oxford, Blackwell.

Davis, M. (1992) 'Beyond *Blade Runner*. Urban control – the ecology of fear', *Open Magazine Pamphlet Series* 23, Westfield, NJ, Open Media.

Deleuze, G. (1986) *Cinema 1: The Movement-Image*, London, Athlone.

Deleuze, G. (1989) *Cinema 2: The Time-Image*, London, Athlone.

Deleuze, G. and Guattari, F. (1988) *A Thousand Plateaus: Capitalism and Schizophrenia*, London, Athlone.

Deleuze, G. and Parnet, C. (1987) *Dialogues*, London, Athlone.

Derrida, J. (1988) *Limited Inc.*, Evanston, Ill., Northwestern University Press.

Derrida, J. (1992) *Given Time, 1: Counterfeit Money*, Chicago, University of Chicago Press.

Derrida, J. (1994) *Specters of Marx: The Work of Mourning, the State of the Debt, and the New International*, London, Routledge.

Dick, P. K. (1972) *Do Androids Dream of Electric Sheep?* London, Grafton.

Doel, M. A. (1992) 'In stalling deconstruction: striking out the postmodern', *Environment and Planning D: Society and Space* 10: 163–79.

Doel, M. A. (1994) 'Deconstruction on the move: from libidinal economy to liminal materialism' *Environment and Planning A* 26 (7): 1041–59

Doel, M. A. (1995) 'Bodies without organs', in S. Pile and N. Thrift (eds) *Mapping the Subject: Geographies of Cultural Transformation*, London, Routledge, 226–40.

Doel, M. A. (1996) 'A hundred thousand lines of flight: a machinic introduction to the nomadic thought and scrumpled geography of Gilles Deleuze and Félix Guattari', *Environment and Planning D: Society and Space* 14 (4): 421–39

Doel, M. A. and Clarke, D. B. (forthcoming) 'Transpolitical urbanism: suburban anomalies and ambient fear' in C. Harrington and G. Crysler (eds) *Street Wars: Space, Power and the City*, Manchester, Manchester University Press.

Freud, S. (1962) *The Ego and the Id*, New York, Norton.

Gasché, R. (1986) *The Tain of the Mirror: Derrida and the Philosophy of Reflection*, Cambridge, Mass., Harvard University Press.

Gibson, W. (1993) 'The Cyberman: interview with William Gibson by John Robb', *City Life* 238: 12–13.

Grist, L. (1992) 'Moving targets and black widows: *film noir* in modern Hollywood', in I. Cameron (ed.) *The Movie Book of Film Noir*, London, Studio Vista, 267–85.

Haraway, D. J. (1985) 'A manifesto for cyborgs: science, technology and socialist feminism in the 1980s', *Socialist Review* 15 (2): 65–108.

Haraway, D. J. (1992) 'When man™ is on the menu', in J. Crary and S. Kwinter (eds) *Incorporations: Zone 6*, New York, Zone Books, 39–43.

Harvey, D. (1989) *The Condition of Postmodernity: An Enquiry into the Origins of Cultural Change*, Oxford, Blackwell.

Instrell, R. (1992) '*Blade Runner*: the economic shaping of a film', in J. Orr and C. Nicholson (eds) *Cinema and Fiction: New Modes of Adapting, 1950–1990*, Edinburgh, Edinburgh University Press, 160–70.

Jameson, F. (1983) 'Postmodernism and consumer society', in H. Foster (ed.) *The Anti-Aesthetic*, Port Townsend, Wash., Bay Press, 111–25.

Jameson, F. (1984) 'Postmodernism, or the cultural logic of late capitalism', *New Left Review* 146: 53–92.

Jameson, F. (1991) *Postmodernism, or, the Cultural Logic of Late Capitalism*, London, Verso.

Kearney, R. (1988) *The Wake of Imagination: Ideas of Creativity in Western Culture*, London, Hutchinson.

Kennedy, H. (1982) 'Ridley Scott interview', *Film Comment* 18 (4): 66.

Kirsch, S. (1995) 'The incredible shrinking world? Technology and the production of space', *Environment and Planning D: Society and Space* 13: 529–55.

Lacan, J. (1977) *Ecrits: A Selection*, tr. Alan Sheridan, London, Tavistock.

Laporte, D. (1978) *Histoire de la Merde*, Paris, Christian Bourgeois.

Latour, B. (1992) 'Where are the missing masses? The sociology of a few mundane artefacts', in W. E. Bijker and J. Law (eds) *Shaping Technology/Building Society: Studies in Sociotechnical Change*, London, MIT, 225–58.

Latour, B. (1993) *We Have Never Been Modern*, Brighton, Harvester Wheatsheaf.

Lefebvre, H. (1991) *The Production of Space*, Oxford, Blackwell.

Lyotard, J.-F. (1984) *The Postmodern Condition: A Report on Knowledge*, Manchester, Manchester University Press.

Lyotard, J.-F. (1988) *The Differend: Phrases in Dispute*, Manchester, Manchester University Press.

Lyotard, J.-F. and Thébaud, J.-L. (1984) *Just Gaming*, Manchester, Manchester University Press.

McCaffery, L. (ed.) (1991) *Storming the Reality Studio: A Casebook of Cyberpunk and Postmodern Science Fiction*, Durham, NC, Duke University Press.

Mandel, E. (1975) *Late Capitalism*, London, Verso.

Marder, E. (1991) '*Blade Runner*'s moving still', *Camera Obscura* 27: 88–107.

Massey, D. (1991) 'Flexible sexism', *Environment and Planning D: Society and Space* 9: 31–57.

Nietzsche, F. (1974) *The Gay Science*, New York, Vintage Books.

Penley, C. (1991) 'Time travel, primal scene and critical dystopia', in C. Penley, E. Lyon, E. Spigel and J. Bergstrom (eds) *Close Encounters: Film, Feminism and Science Fiction*, Minneapolis, University of Minnesota Press, 63–80.

Riviere, J. (1929) 'Womanliness as a masquerade', *International Journal of Psychoanalysis* 10: 303–13.

Salisbury, M. (1992) 'Back to the future', *Empire* 42 (December): 90–6.

Silverman, K. (1991) 'Back to the future', *Camera Obscura* 27: 108–133.

Spivak, G. C. (1983) 'Displacement and the discourse of woman', in M. Krupnik (ed.) *Displacement: Derrida and After*, Bloomington, Indiana University Press, 169–90.

Thompson, M. (1979) *Rubbish Theory: The Creation and Destruction of Value*, Oxford, Oxford University Press.

Wakefield, N. (1990) *Postmodernism: The Twilight of the Real*, London, Pluto.

Wollen, P. (1991) 'Film aesthetics, film history and the idea of a film canon', paper delivered to the Columbia Film Seminar, 26 September, 1991, University of Missouri, Columbia, Missouri.

Žižek, S. (ed.) (1992) *Everything You Always Wanted to Know About Lacan: But Were Afraid to Ask Hitchcock*, London, Verso.

8

'THE PEOPLE IN PARENTHESES'

space under pressure in the post-modern city

•

Elisabeth Mahoney

all spatialities are political because they are the (covert) medium and (disguised)
expression of asymmetrical relations of power.
KEITH AND PILE, *1993: 220*

Space, then, has a history ... It is a product of representations.
BURGIN, *1990: 108*

To speak of the urban, to describe the cityscape, immediately brings spatial paradigms and metaphors into discourse. Our literal and figurative ways of seeing and
naming the city are grounded in the spatial; fantasies, theories and architectural
projections of future cities foreground space over place and time: Mike Featherstone
describes the city of postmodernity, for example, as 'a no-place space' (Featherstone
1991: 99); Edward Soja, in his reading of postmodern Los Angeles, insists upon
'a geography of simultaneous relations and meanings that are tied together by a
spatial rather than temporal logic' (Soja 1989: 1). In this chapter, my focus is on
the politics of such spatial categories, narratives and metaphors and, in particular,
the gender politics of urban space. To what extent, and in what ways, can we
trace connections between the spatial and sexual politics of a text? My discussion
of three recent city-films will read across the texts to look at 'relations of power'
as inscribed in and through urban space, and to recover spatial configurations from
their conventional place somewhere outside of, or beyond, linear, hierarchical, ideological, and narrative structures.

The cities of postmodernity are obviously suggestive places for thinking through
connections between spatial categories and these dominant structures: the postmodern city has been conventionally theorized as a site of difference, fragmentation,
conflict, and plurality, and, as the comments by Soja and Featherstone demonstrate,
it has also tended to be represented in spatial rather than temporal terms. I am
interested also in extending feminist cultural studies of city-representations into the

postmodern. There has been a substantial amount of feminist work on gender and the cities of modernity;[1] the grounding of the modernist crisis of identity through and in the urban *and* the feminine; and the importance of discourses on the urban in shaping and reflecting dominant ideologies of sexual difference and femininity.[2] If, as many theorists have claimed, the postmodern is marked by a profoundly different experience of the urban, does this produce different configurations of urban space in theories and fictions? Crucially, does this challenge or put under pressure conventional narratives of sexual difference as mapped out in the cityscape?

The rhetoric of postmodern theorizing on the urban would suggest that this is the case. Writing on the importance of imaginary cities in our experience of the urban landscape, Paul Patton locates the postmodern paradigm shift specifically in the cityscape:

> To the extent that the inhabitant of the (post)modern city is no longer a subject apart from his or her performances, the border between self and city has become fluid ... the city as experienced by a subject which is itself the product of urban experience, a decentred subject which can neither fully identify with nor fully dissociate from the signs which constitute the city.
>
> (Patton 1995: 117–8)

'His or her' subjectivity in the postmodern is 'the product of urban experience'; it is performative, decentred and fluid, the result of a rupture of the relations between identity and the urban. In *Postmodern Geographies*, Edward Soja argues that this transformation in our experience of the urban in postmodernity directly challenges existing critical practices of reading the urban which, he suggests, have historically privileged the temporal, linear and historical over the spatial. His aim is to 'spatialize [this] historical narrative' (Soja 1989: 1) and, to do so, Soja insists that textual identities – the shape, organization and emphases of theoretical, critical texts – must change to reflect experience of the postmodern cityscape. Commenting on his own text, he declares: 'The perspectives explored are purposeful, eclectic, fragmentary, incomplete, and frequently contradictory, but so too is LA ...' (247).

A different lived experience of the urban, it seems, calls for a different theoretical paradigm and critical practice. Theoretical space, as much as any other, is under pressure to jettison linearity and the desire for panoptic vision; to bring otherness, difference and the eclectic into view. However, as feminist critics working in urban and cultural studies have argued, in both positive and negative critiques of postmodernism, there has been a marked resistance to anything more than a gesture towards multiplicity in the critical and/or urban space. This is Doreen Massey's critique of both Soja and David Harvey's accounts of 'the postmodern condition' (Massey 1991). While both read postmodernism as a space of contestation and eclecticism, and as the result of economic and political paradigm shifts, both ground their critiques, she argues, in hyper-conventional representations of 'otherness'.

Massey focuses on the gulf between the rhetoric of difference and, in practice, the 'continued subordination for all those people in parentheses – those who do not in their complex identities match the postulated, uncomplicated – because unanalyzed – universal' (Massey 1991: 55). If we look at Soja's *Postmodern Geographies*, we can see this clash between, on the one hand, his concern with difference, otherness and critical 'ghettoization', and, on the other, the continuation of such 'ghettoization' within the text. Having suggested that his text's identity mirrors that of the city it describes, he continues the spatial metaphor in claiming that *Postmodern Geographies* is shaped like 'a map' (Soja 1989: 1); specifically, a map which will not exclude and peripheralize:

> This reconstituted critical human geography must be attuned to the emancipatory struggles of all those who are peripheralized and oppressed by the specific geography of capitalism (and existing socialism as well) – exploited workers, tyrannized people, dominated women.
>
> (Soja 1989: 74)

And yet, if we look closely at this textual space, beyond the rhetoric of difference, we find conventional relations between women, cities and theory: in Soja's index there is only one reference to 'feminism' and none to women. So what does 'attuned' mean? Does naming 'otherness' in one's text suffice? Is self-conscious, self-reflexive objectification of 'tyrannized' people some kind of radical shift? Clearly, the 'specific geography' within which certain experiences of urban life are kept as marginal, discursively placed on the city limits, is still in place here. This is in part due to Soja's rhetoric and practice of mastery over the urban spectacle, as evidenced by two contrasting chapter titles within his text: 'It All Comes Together in Los Angeles' (mainly economic analysis) and 'Taking Los Angeles Apart: Towards a Postmodern Geography' (a 'succession of fragmentary glimpses'). Mastery of the cityscape is explicit here on two levels. First, the 'fragmentation' is only of one perspective, the 'glimpses' are all from the same critical position; second, despite this fragmentation of the critical view, the postmodern theorist is able to construct and deconstruct the city-as-text. The city, while marked as chaotic and a site of otherness, is mastered from the critical space: the distance between the two is not breached here. Commenting on the spatial form, and consequences, of the topography of Los Angeles, Soja unwittingly provides an ironic analogy for his own mapping of the city-as-text, and the otherness contained within it:

> Underneath this semiotic blanket there remains an economic order . . . an essentially spatial division of labour . . . this conservative deconstruction is accompanied by a numbing depoliticization of fundamental class and gender relations and conflicts. When all that is seen is so fragmented and filled with whimsy and pastiche, the hard edges of the capitalist, racist and patriarchal landscape seem to disappear, melt into air.
>
> (Soja 1989: 246)

This last image – the illusion of a depoliticized, 'soft' postmodern landscape (lived or metaphorical) which in fact veils persistent and very real oppressive ideologies – has been the starting-point for feminist critiques of postmodern theory.[3] With particular regard to urban studies, feminists have argued that beneath the façade of 'whimsy and pastiche' of the postmodern city, lie entrenched narratives of sexual difference.[4] Such feminist critiques of modernist and postmodernist theories and representations of urban space suggest that the ways in which cities have conventionally been theorized and imagined are both shaped by, and reflect, wider social and political discourses. In *Feminism and Geography*, for example, Gillian Rose interrogates ideologies of space and gender in geography, arguing that dominant discourses on public space are inextricably linked to consensual, conventional socio-political formations. She contrasts the traditional coding of public space as masculine (which implies and encodes the invisibility of women in the urban, their presence always problematic and transgressive) with feminist contestation of such space, in theory and practice. Rose uses the example of Take Back the Night marches to argue that 'a new claim on public space' implies 'a new social form' (Jones 1990: 803; quoted in Rose 1993: 36), linking such demonstrations about the dangers and limits of the cityscape for women with theoretical uses of spatial images in recent feminist theory. Recent emphasis on a possible 'politics of location' for women, on the 'blindspots' and 'elsewhere' of discourse, demonstrates the politicizing and gendering of spatial categories, conventionally read – like all philosophical and cultural 'spaces' – as universal, beyond ideology. Like Soja, Rose insists that the changing nature of urban experience calls for new critical paradigms, but argues that these have yet to emerge in mainstream human and urban geography. This is, she argues, because of a reluctance to give up the position of mastery of the urban spectacle and of the objects contained within it:

> Masculine geographers are by and large demanding an omniscient view, a transparent city, total knowledge . . . feminist geographers are understanding the contemporary city not as the increasing fragmentation of a still-coherent whole, but rather in terms of a challenge to that omniscient vision and its exclusions.
>
> (Rose 1993: 133)

In recent work on the city, feminist philosopher Elizabeth Grosz has offered a theoretical 'challenge' to the exclusions contained within conventional discourses on the urban. Her work, while not directly concerned with aesthetic representations of the city, is important for my reading of filmic texts later in the chapter and I want to sketch out here the approach Grosz takes in general terms. Her project has been to trace the philosophical and spatial foundations of our ways of interpreting the city, and to read these specifically in terms of gender. Like Rose, she argues that the organization, representation and theorizing of the urban is shaped by, and works to shape, wider socio-political formations; in other words, that cities can be read as sites and networks of power and, importantly, that these shift and change in different historical and cultural matrices. Conventionally, Grosz argues, the city has been read as a static and non-ideological space, and to demonstrate this she outlines two traditional paradigms of reading the city, both of which invoke particular representations of corporeality. First, Grosz describes a paradigm which she labels humanist: here, the city is pre-dated by the body, and develops according to changes in human needs and patterns of settlement: 'the human subject is conceived as a sovereign and self-given agent which, individually or collectively, is responsible for all social and historical production. Humans *make* cities' (Grosz 1992: 245). This model of the city is problematic, Grosz suggests, because it allows for only a one-way relationship between body and city, and also because it conceives of the 'human' in terms which privilege the mind over the body. The second model posits an isomorphic relationship between the city and the body. This can be most clearly seen in the metaphorization of the body in political discourses on the State: for example, the notion of the 'body politic'. Grosz critiques this tradition as phallocentric (defining phallocentrism 'not so much the dominance of the phallus as the pervasive unacknowledged use of the male or masculine to represent the human'; Grosz 1992: 247) in its coding of the body and city/public space/State as masculine.[5] In place of these traditions, Grosz delineates a new paradigm for reading the city and this necessarily involves a shake-up in the relational representation of the body:

> [T]here is a two-way linkage which could be defined as an *interface*, perhaps even a co-building. What I am suggesting is a model of the relations between bodies and cities which sees them, not as megalithic total entities, distinct identities, but as assemblages or collections of parts, capable of crossing the thresholds between substances to form linkages, machines, provisional and often temporary sub- or microgroupings.
>
> (Grosz 1992: 248)

Here we see in practice the delineation of new theoretical spaces and spatial metaphors for viewing the city: here bodies and the urban meet at an interface in transitory, fleeting moments of connection. Replacing the humanist and isomorphic models of body–city connection with mere 'provisional . . . linkages' jettisons the possibility of any mastery over the urban spectacle from one fixed viewpoint.

While this new paradigm brings with it an implicit challenge to existing critical practices in urban studies, Grosz's more recent work on gender and space presents, I believe, a much more urgent and explicit critique of traditional ways of reading the cityscape. Grosz interrogates the spatial entity of the *chora* (a pre-social, feminine and, specifically, maternal space which she describes as 'the locus of nurturance in the transition for the emergence of matter, a kind of womb . . . the space in which place is made possible, the chasm for the passage of spaceless forms into a spatialized reality'; Grosz 1995: 50–1) in the writings of Plato and Derrida, arguing that it offers an example of the coding of feminine spaces – and spaces of femininity – as enigmatic, silent and the negative Other, but also as a 'support or precondition' of the masculine and social (Grosz 1995: 50). Thus, underlying Platonic and Derridean readings of the city and architecture we find a common space: the maternal and corporeal realm of the *chora* in which their theories of place are grounded. Grosz extends this example to argue that conventional models of urban place and space demonstrate this suppression of the feminine and bodily on the one hand, and the metaphorization of the feminine to symbolize the pleasures and dangers of the cityscape on the other:[6]

> their various philosophical models and frameworks depend on the resources and characteristics of a femininity which has been disinvested of its connections with the female, and especially the maternal, body and made to carry the burden of what it is that men cannot explain, cannot articulate or know, that unnameable recalcitrance that men continue to represent as an abyss, as unfathomable, lacking, enigmatic, veiled, seductive, voracious, dangerous and disruptive but without a place or name.
>
> (Grosz 1995: 57)

Grosz thus uses the specific entity of the *chora* to trace a philosophical, representational placing of the feminine which, I want to suggest in my readings of recent filmic city-texts, describes the conventional spatial positioning of women across a wide range of urban discourses. The persistence and currency of these contradictory representations of femininity in (and as) the city – 'an abyss', 'lacking', 'veiled', 'voracious' – raises questions as to how spatial formations might be subverted, how dominant spatio-temporal relations (which clearly encode other power relations) might be disrupted. Of particular interest here, in my reading of gender and space in three film-texts, is the extent to which space might be subverted within the framework of non-avant garde, narrative, 'realist' film where we might expect to find a reliance upon conventional spatial paradigms. In the textual discussion which follows, I begin by reading two film texts, Joel Schumacher's *Falling Down* (1992) and Jim Jarmusch's *Night on Earth* (1992), which represent, in very different ways, the postmodern city, arguing that in each case, the city of eclecticism, difference and 'otherness' screens or veils hyper-conventional figuring of urban space and the feminine. It seems crucial to me to uncover such spatial formations in contemporary as well as in historical texts and to remind ourselves that it is not only in, for example, discourses relating to the cities of modernity that we find the space (feminine)/ place (masculine) opposition described by Grosz. Foregrounding such a structuring opposition – which, as Grosz demonstrates, is so embedded in our conception of the urban that it remains almost unseen and unspoken – in texts is a first stage in the disruption of this convention. I then move on to discuss a third text, Leslie Harris's *Just Another Girl on the I.R.T.* (1992), which also represents the contemporary city and has a conventional narrative structure but which, I argue, subverts the conventional paradigm of gender and urban space.

My first two texts offer one negative and one more positive representation of the contemporary, postmodern city, and there are other important differences of context. *Falling Down*, a mainstream Hollywood film, went on general release with the star attraction of Michael Douglas, emerging out of, or at least aspiring to, a context of 'State of the Nation' films; at the time of release the film was repeatedly described as 'a *Taxi Driver* for the nineties'. *Night on Earth* can be placed in two different contexts: obviously, it is an art-house movie, but the film also has connections with other texts such as Alan Rudolph's *Equinox* (1992) and Wim Wenders' *Until the End of the World* (1991); in each of these, the consequences of technological and cultural shifts in contemporary urban life are explored. In the three films, the homogeneity and simultaneity of urban experience is a central concern and, in *Night on Earth* at least, is celebrated.

Falling Down represents the dislocation between subject and city described by Paul Patton in his comment on the postmodern cityscape: the city here encodes absolute alienation and fragmentation and focuses particularly on a crisis of masculinity through and in the urban. The narrative charts the journeys of two male characters: William Foster (Douglas) and a retiring police officer, Prendergast (Robert Duvall). Both are trying to 'go home'; Prendergast to his paranoiac wife who has forced him to retire and move out of the city, Foster to the house he used to share with his now ex-wife, Elizabeth (Barbara Hershey), and child. Foster's journey, which is signalled from the beginning of the film as an impossible, nostalgic fantasy, is also clearly positioned as a pilgrimage, against the odds, by an 'Everyman' figure. Rather like Winston Smith in *Nineteen Eighty-Four* (1948), which George Orwell had initially wanted to call *The Last Man in Europe*, Foster – with his 'D-FENS' number plate – is represented, albeit *in extremis*, as the last 'd-fender' of an old cultural and social order which has long since disappeared. Indeed, his real name is replaced by 'D-Fens' in the film's credits, as if to signal that by the end of the narrative, his former identity has been completely erased. As Carol Clover argues, Foster represents a now obsolete universality (read masculinity), the centre which no longer holds: 'It is the great unmarked or default category of Western culture, the one that never needed to define itself, the standard against which other categories have calculated their differences' (Clover 1993: 9). This obsolescence is signalled by a particular construction of urban space, as hostile and violent, but also unreadable, out of control: the cityscape represents, contains and triggers Foster's violent journey. Los Angeles here is a site of stasis, the city is rotting; the film's opening sequence constructs public space as claustrophobic, invasive and stagnant (the seemingly endless traffic queues, the details of everyday life which, in unavoidable close-up – the Garfield toy, the fly trapped in Foster's car, the car sticker in front: 'How Am I Driving? Call 1-800-EAT SHIT' – become signs of an alien landscape). Foster has no sense of ownership or rootedness; the space of the city has been taken over by racial and/or sexual 'others'.

It is not only public space within which Foster is dispossessed and obsolete, however. The narrative constantly juxtaposes two spaces to mark his alienation: the cityscape is contrasted throughout with the seemingly safe, maternal, feminized space of 'home'. The use of juxtaposition as a structuring, narrative device in the text is particularly striking in a sequence which begins as Foster calls his ex-wife

8.1 *Falling Down*
© Warner Bros. Courtesy of BFI Stills, Posters and Designs.

from a public pay-phone, just before a drive-by shooting occurs. By this point in the film, the protagonist's descent into violence is well under way. The tension of this sequence is produced by two layers of out-of-control, masculine violence: the gang's desire for revenge for an earlier humiliation on their territory (Plate 8.1) which leads them (once the only woman in the car has been forcibly removed) to shoot openly and randomly in daylight in a crowded street; and Foster's insistence that he is 'coming home' and has to bring his daughter a birthday present, despite a restraining order barring him from the house. The scene is constructed to show the violence ('real' and epistemological) that is *done* to Foster, however. His very lack of place in both the urban and the domestic spaces is marked at the same time: just as the gang spot Foster in the street, his ex-wife tells him, 'This isn't your home any more'.

For all his violence and extremity, we are supposed to find Foster an empathetic character, to understand his frustration with, if not his response to, the violence done to him by lived experience. To construct this, the spatial juxtaposition continues to structure the narrative. After the drive-by shooting, Foster is able to walk calmly through the carnage to the gang's car, which has been involved in a crash, to assert his mastery, as he does throughout the film, in desperate performances of masculinity through violence (Plate 8.2). Here, he greets the gang-leader with 'You missed' and, after shooting him, advises 'Get some shooting lessons,

8.2   *Falling Down*
© Warner Bros. Courtesy of BFI Stills, Posters and Designs.

asshole'. This is an important scene as it uses humour to detract from the act of violence (which could not be more differently represented from that of the drive-by shooting, shown to be irrational and, subsequently, unsuccessful) and, through the use of slow motion, apocalyptic music, and the contrast of Foster's still brilliant white shirt with the assault of colour on the sultry, heat-hazy street, marks *Foster* as 'other': as unreal, as out of this corrupt and crazed world.

To reinforce this, the sequence continues with another important juxtapositioning of public and domestic space. Two scenes in a public park are contrasted with a police interview at the house of Foster's ex-wife. It is here that the interconnect-edness of the two spaces becomes explicit and the text begins to give us clues as to the root causes of Foster's displacement. In the first scene in the park, Foster tries, unsuccessfully, to board a bus which is surrounded by a large crowd. We see his impression of the public space: once again as claustrophobic and unreadable, but here the contrast with the safe space of home is further underlined by a poster featuring a small girl, with tears rolling down her cheeks, with the words 'I love you daddy' underneath. The incommensurability of the two spaces is made explicit here: Foster is unable to make his way through the crowd to get on the bus to continue his journey home. He remains in the public space which, in the second sequence, is even more carnivalesque: groups of buskers, beggars (one holding a sign which reads 'We are dying of AIDS'), the mad and the lonely. A fight breaks out between a group of street people; the music, colour and heat assault Foster's senses and, in the middle of the space is a sign which reads 'NO MATTER/NEVER MIND'. This sign, of course, is the antithesis of the poster of the young girl: in the second sequence, the lack of connection, communication and relationship is being accepted, if not celebrated.

These two images of public space are juxtaposed with the scene inside the home of Foster's ex-wife which is important in that it calls into question the woman's judgement of her husband as dangerous and, thus, suggests that Foster's displacement has been *caused* by a misunderstanding of him within the domestic, feminized space of home. The ex-wife is represented as scatty (she is unsure if the restraining order is a hundred yards or metres) and neurotic. Although a judge has placed an order on Foster, the police officer is incredulous as Beth explains that her husband had never struck her or their daughter and that her fear is based on strong suspicion that Foster could turn violent. As the officer blurts out Not *exactly*? and You *think*? to her responses to his questions, the relationship between the two spaces in the film shifts. Suddenly it is possible to see the 'power' located in the feminized, domestic space and not (by implication, not any more) in the public realm. At this moment, the text reveals itself to be in many ways a 'backlash' film and to be laying blame for the 'State of the Nation' firmly in this private space: Foster, and the default category he represents, have been made redundant within it.

This sexualizing of space works throughout the film and is not only connected to Foster and his ex-wife. His mother occupies only private space: she is cut off from external reality, knows little or nothing about her son's life and comes alive only when talking about her collection of glass animals. The fact that we see Foster's spaces of home – of both his childhood and his marriage – is crucial: again the film links his excess and inability to survive in the public world to the spaces which he formerly inhabited. This pattern is repeated in the narrative around the retiring police officer, Prendergast. He is shown to be withdrawing from public life, and policing the city, because of his neurotic wife. Like Foster's mother, she has made the space of the home into an alienating one: she is obsessive, seemingly afraid to leave the house, and childlike (Prendergast has to sing 'London bridge is falling down' to quieten her). Both male characters have only one 'safe' space and that is memory, presumably when their place in the domestic space was assured and unquestioned: we see short excerpts of family videos shot by Foster in happier times (although the strains are clearly there) and the photograph of Prendergast's daughter who died at the age of two. Prendergast is able to emerge from this journey to claim a new place in the urban space (at the end of the film he is not going to retire) as he is able to reinstate his position as 'master' of the domestic space (his wife becomes acquiescent, apologetic, wifely, and not-mad when he shouts at her to 'shut up' and tells her to have his dinner ready for him when he comes home). Foster is unable to transform his position in this way and dies unable to understand his place as villain in either of the spaces of the film: 'I'm the *bad* guy? How'd that happen?'

Juxtaposition is also an important structuring device in *Night on Earth* as Jarmusch records one moment, in one night, in five cities: Los Angeles, New York,

Rome, Paris and Helsinki. In each of the cities, the narrative unfolds in a taxi; this both captures the transitory nature of urban experience – unpredictable meetings and journeys – and foregrounds what is usually deemed insignificant, trivial, everyday, as central to this film. In interview, Jarmusch emphasizes this point:

> So in a way the content of this film is made up of things that would usually be taken out. It's similar to what I like about *Stranger than Paradise* or *Down By Law*, the moments between what we think of as significant.
>
> (Jarmusch 1991: 9)

This attention to 'things that would usually be taken out' signals a concern with at least the details of realism. Commenting on the different registers of French used in the Paris scene, for example, Jarmusch argues 'these things are important so that it feels real' (*ibid.*). This is a striking comment from a director not conventionally championed as a realist, whose film works to construct and deconstruct stereotypes as a narrative device. In each of the five narratives, national, regional, class and gender stereotypes are reproduced and restylized in these arbitrary connections in urban landscapes. This is of course a postmodern and playful use of such stereotypes; Jarmusch is positing identity as ironic and performative, and deliberately sets out to defeat expectations of such stereotypes. Seemingly, the urban space, in this positive, self-conscious and playful text, offers a place for the deconstruction of conventional ideologies of racial or sexual difference.

I want to suggest, however, that in *Night on Earth*, while we see a reconfiguration of urban space and a release from some traditional discourses on the attraction and repulsion of the city, for example, we also see a reliance on stereotypes of gender and race, which means that it occupies, like *Falling Down*, a very conventional place in terms of its representations of sexual difference. This conventionality is once again grounded in points of connection between urban and gender ideologies. Narrative might be liberated in this film by the play with time and location, but as we see in the New York and Paris sections in particular, women are reassigned to a traditional (here, transgressive) place in the city. In the New York narrative Angela (Rosie Perez) is a feisty, fast-talking and uncontrollable woman who very much comes to symbolize Brooklyn for Helmut, the taxi driver from Eastern Europe. In the Paris section of the film, one of the passengers is a blind young woman (Beatrice Dalle) who takes a taxi driven by an African driver. In both narratives, despite the parodic treatment of other characters and situations, these female characters occupy very conventional discursive spaces. Angela is portrayed as wild and exotic, but also, for Helmut, as enigmatic and elusive in the same way – and at the same time – as New York is: he describes both Angela and Brooklyn Bridge as 'beautiful', with the same facial expression of wonderment and innocent fascination for both. Whether intentionally or not, Jarmusch has reproduced a series of stereotypes around this female character which do not work as parodic in the way they do with other characters in the film (such as Yoyo, Angela's brother-in-law, or Helmut); some stereotypes are clearly more resistant to postmodern restyling than others.

This is also true of Dalle's character in the Paris section. The theme of this narrative is simple enough: although the character is blind, she 'sees' more than others in her narrative and the rest of the film. The parody of this scene is connected to the power that the woman has over the taxi driver despite her blindness. She can tell exactly the route he is taking, she can trace his accent, and she rejects all of his ideas about her life as laughable. She is also shown to be aware of his gaze in the driver's mirror as she freshens her make up and adjusts her bra strap. The humour here is supposed to be in the fact that the driver is unable to avert his gaze, and *she knows this*, particularly as she tells him of the sensual pleasures of love-making without sight. This all takes place in the taxi, on the deserted outskirts of town, in the middle of the night. Two things break down the parodic distance here and they are both connected with gender and the urban space. First, thinking back to Jarmusch's comment that the film should 'feel real', it is ironic that it is actually the specific 'real' represented here which is problematic. As a spectator, I found myself immediately thinking of the dangers that the scene in Paris did not address, the very real limitations on women in the city, particularly at night. Second, the playfulness and irony of the text is at least compromised by the use of a sexualized, glamorous, sensual woman to represent any city, but especially Paris (Plate 8.3). This is of course highly conventional and, seemingly, non-parodic here and complicates any response to my first point on the grounds that Jarmusch is simply reproducing images, signs and stereotypes in a parodic fashion.

Certainly, the critical reception of these two female characters shows that they can be read as reproducing certain dominant narratives of feminine sexuality and women in cities. In J. Hoberman's review of *Night on Earth* he describes Angela as follows: 'Perez's patented head-bob – emphasizing the fastest mouth in American movies – generates enough exoticism to warm the wide-eyed driver (not to mention Jarmusch)' (Hoberman 1991: 8). The conflation of actress and character is striking enough, but it is the conflation of black actress/character and 'exoticism' that is most revealing: this exoticism, this otherness, is clearly marked as sexual and readable by any male spectator ('to warm the wide-eyed driver . . . Jarmusch' and presumably the critic as well).[7]

The discourse reserved for the blind woman in the Paris scene marks her much more clearly as sexual, transgressive and monstrous: 'Eyes rolling in her head, head swivelling on her neck, Dalle is beautiful and monstrous' (Hoberman 1991: 8); Dalle is introduced in his review as 'the eponymous self-mutilated heroine of *Betty Blue*' (*ibid.*). To score the urban landscape as other, it seems, Jarmusch has to reproduce and proliferate images of woman as excess in the city; these are then focused on by the critic to detail the text's otherness. For example, Hoberman, attempting to describe adequately the narrative fluidity of the film, comments on the 'endless loops' in the film, as 'one immigrant driver [melts] into another as Rosie Perez mutates into Beatrice Dalle' (*ibid.*). Although no one has a secure, safe place in Jarmusch's city (and there is no desire or nostalgia for such a place) and there is no sense of ownership of the cityscape, hence the concentration on immigrant identity, women are doubly displaced: monstrous, linguistically or sexually

8.3   *Night on Earth*
© Electric Pictures. Courtesy of BFI Stills, Posters and Designs.

excessive, 'mutations' in the urban space. Just as in *Falling Down* where the crisis in identity in the city was revealed to be grounded in a feminized, private space, in *Night on Earth* the playful representation of the simultaneity of urban experience is grounded in certain points of racial and sexual fixity which have conventionally framed readings of the city space.

In the discussion of my third text, I want to suggest that it is possible to free a text from such points of fixity even within a realist, narrative framework, and to challenge traditional sexualizing of space in terms of gender. Leslie Harris's *Just Another Girl on the I.R.T.* (1992) is, in many ways, an extremely conventional narrative: essentially a film about teenage angst; maturation; and love. In the film, Chantal (Ariyan A. Johnson) is a bright young student from the 'projects' of Brooklyn with serious ambitions to be a doctor (Plate 8.4). She falls pregnant, but refuses to accept the reality of the situation for a long time (she takes a friend on a shopping spree with the money her boyfriend borrows for an abortion). Finally she has the baby and ends up a year later at college, somewhat behind in her plans, but back on track. Clearly, the most important aspect of the narrative is the 'survival against the odds' and some critics have focused on, for example, the seemingly pervasive materialism displayed by Chantal and her friends as a negative part of the film. My interest in the text is obviously more concerned with the

8.4 *Just Another Girl on the I.R.T.*
Courtesy of BFI Stills, Posters and Designs.

relations between gender and the city, sexuality and space. In this regard, Harris has produced a text which challenges the exclusions and stereotypical representations in *Falling Down* and *Night on Earth*.

Like the films of Jarmusch and Schumacher, this text also uses the journey motif, but it is a journey through the late teens of a young black woman, set entirely in the 'projects' she has grown up in, which offer her both something to escape and something to value: the domestic space of the family home is claustrophobic, but gives a sense of rootedness in the city which Schumacher and Jarmusch do not attempt to inscribe. The film has a circular narrative, opening with a scene in darkness in which a man dumps something in a rubbish bag on the street. This, we realize at the end of the film, is Chantal's baby which she initially rejects and then soon rescues. The voice-over here is Chantal's and she begins her story immediately with a sense of belonging in urban space: 'Some people hear about my neighbourhood and assume some real fucked-up things, but I'm gonna tell you all the real deal.' From the sequence which accompanies the opening credits of the film, the 'projects' provide public spaces which Chantal and her female friends regard as their own: for example, there is a scene where three young women on a park bench discuss contraception and the risks of unprotected sex. Also, the long sequence with the opening credits shows Chantal travelling across town and partly because she narrates to camera and partly because of the movement and music of the scene, there is a sense of ownership, articulation and visibility which is extremely rare in representations of women in urban space. The cityscape is opened up to include other voices and experiences, as is signalled by the text's title. Rather than the 'Everyman' figure of *Falling Down*, we have 'just another girl'; rather than apocalyptic unreadability, we have an emphasis on the everyday (the chorus of the track which accompanies the opening credits is: 'It's just my life, yeah, and it's just all right'). Most importantly, in place of the exoticized and/or eroticized woman used to symbolize urban space in Jarmusch's film we have a protagonist who, through her first-person narration and *place* in the city, refuses the position of objectification occupied by the characters played by Dalle and Perez.

The film explicitly raises the issue of spatial politics, gender and race, in two scenes set in Chantal's history lesson at school. In the first, Chantal questions what she sees as the Eurocentrism of their history classes, arguing that the black urban students in the room know little or nothing about their own history: 'we need to know our history to live,' she insists. Rather than learning the history of other cultures, she asks for lessons on the social and political conditions which have produced such inequalities in American life, especially contemporary urban life. In the second scene, the pedagogic space has been transformed, drawing attention to the power structures contained within it: Chantal is seated at the teacher's desk and the teacher, the only white in the room, is sitting in the same area as his students. Chantal gives an alternative history lesson which, interestingly, is more of a geography lesson: in place of linear, temporal history, Chantal uses maps to focus her revision on spatial categories. She illustrates that maps have traditionally diminished the size of certain parts of the world and have exaggerated the space occupied by, for example, Europe and North America. When the teacher challenges her information, she gives him and the class the name of an African–American academic woman who is quite willing to come and talk to the students; the teacher acquiesces and Chantal is cheered by her classmates. In these scenes, she marks spatial configurations as always politicized; as ideological; gendered; oppressive; disembodied; and hierarchical but, importantly, as reclaimable. Chantal's relationship with the urban space of the 'projects' throughout the film enables her to work through larger questions of power and territory in both public and private space. Most importantly, the space which she occupies in the city and in the text is not a marginal or silent space; she is not the 'other' against whom the urban spectator defines 'himself', but rather, through her act of self-representation in the city, Chantal offers up a direct challenge to the traditional cultural positioning of women as other, as metaphor, as spatial ground, in masculine experience and appropriation of public space.

In my reading of these three films, all released in the same year and all dealing with the contemporary, postmodern American city, I have focused on spatial configurations as represented (usually implicitly) in a network of gender relations. In both *Falling Down* and *Night on Earth*, despite marked differences in the aesthetic and political projects framing the texts, there is a reliance on the same, conventional representation of this network. In arguing that Leslie Harris's film radically

challenges such conventions, I do not intend to claim any special, essential difference in women's experience of urban space or in the way women directors, for
example, might choose to represent the cityscape. To claim a special relationship
between women and the spatial is to lock ourselves back into the philosophical
set of binary oppositions described by Elizabeth Grosz, in which women can never
claim a social or cultural *place*. However, I would argue that texts which set out
to inscribe experiences of the city previously marginalized or marked as 'other' do
implicitly or explicitly challenge traditional discourses on urban space. Mathieu
Kassovitz's *La Haine* (1995), for example, shows the fiercely guarded space of the
metropolitan centre of Paris under threat from its 'others' placed out in housing
schemes on the city's outskirts. Here, urban space is always contested and a site
of potential conflict; through violence and resistance, the appropriation of this
space by the centre is made explicit. Both *La Haine* and *Just Another Girl on the
I.R.T.* represent the possibility of reclaiming or re-imagining the space of the city
in the postmodern, a possibility which Donatella Mazzoleni insists we must explore
if we are to produce 'a new imaginary' to match our changing experience of the
urban, and which, I have argued in my readings of *Falling Down* and *Night on
Earth*, has yet to emerge in mainstream American cinema:

> We must note, at this point, a different urgency . . . the urgency of confronting
> collective living right where its most radical possibility of crisis exists: that
> is, in its imaginative roots. The destruction which comes from the outside
> (even from earthquakes) seems to us less serious than the destruction which
> comes from within if we cannot learn to live in our post-metropolitan condi-
> tion, to live beyond the uprooting of the affinity between the I and the world
> by positing a new imaginary.
>
> (Mazzoleni 1990: 91)

NOTES

1 See Pollock (1988), Wilson (1991) and
Wolff (1985, 1994).

2 Such as Judith Walkowitz's study of
late-nineteenth century London, *City of
Dreadful Delight* (1992); Patrice Petro's
work on representations of femininity and
urban decadence in Weimar Germany
(1989), and Constance Balides' reading
of the sexualizing of the feminine body
in 'everyday life' film texts (e.g. street
films) in pre-1907 American cinema
(1993).

3 See, for example, Morris (1988),
Nicholson (1990) and Finn (1993).

4 See Mahoney (1994).

5 I am summarizing Grosz's argument
here. See Grosz 1992: 244–9.

6 One of the most explicit examples of
this metaphorization of the urban through
the feminine is Harold Nicholson's description of the differences between European
cities. This represents the figuring of the

feminine outlined by Grosz: 'London is an old lady in black lace and diamonds who guards her secrets with dignity and to whom one would not tell those secrets of which one was ashamed. Paris is a woman in the prime of life to whom one would only tell those secrets which one desires to be repeated. But Berlin is a girl in a pull-over, not much powder on her face, Hölderlin in her pocket, thighs like those of Atlanta, an undigested education, a heart which is almost too ready to sympathize . . . Berlin stimulates like arsenic, and then when one's nerves are all ajingle, she comes with her hot milk of human kindness; and, in the end, for an hour and a half, one is able, gratefully to go to sleep' (Nicholson, quoted in Petro 1989: 40).

7 This reading of racial 'otherness' in terms of exoticism within a playful, postmodern text alerts us to the persistence of metanarratives, marginalization and objectification not only within the primary text itself, but in ways of looking and reading. Hoberman's comments on the character of Angela made me think of Baudrillard's description of some of the urban communities in *America*. In this text, the theorist perhaps most closely aligned with postmodernity tries to describe the 'otherness' of women of colour that he sees on the streets of New York: 'The beauty of the Black and Puerto Rican women of New York . . . Apart from the sexual stimulation produced by the crowding together of so many races, it must be said that black, the pigmentation of the black races, is like a natural make-up that is set off by the artificial kind to produce a beauty which is not sexual, but sublime and animal – a beauty which the pale faces so desperately lack' (Baudrillard 1988: 15–16). 'Sublime and animal'; the 'sexual stimulation produced by the crowd': Baudrillard's text – and view of the urban – is obviously framed by the kind of conventional binary oppositions and *flâneur*-like impressions of the city as texts describing the cities of modernity. The position of mastery and distance has not been called into question.

FILMS CITED

Falling Down (d. J. Schumacher, 1992)

Night on Earth (d. J. Jarmusch, 1992)

Just Another Girl on the I.R.T. (d. L. Harris, 1992)

Equinox (d. A. Rudolph, 1992)

Until the End of the World (d. W. Wenders, 1991)

La Haine (d. M. Kassovitz, 1995)

REFERENCES

Balides, C. (1993) 'Scenarios of Exposure in the Practice of Everyday Life: Women in the Cinema of Attraction', *Screen* 34 (1): 19–37.

Baudrillard, J. (1988) *America*, London: Verso.

Burgin, V. (1990) 'Geometry and Abjection', in J. Fletcher and A. Benjamin (eds) *Abjection, Melancholia and Love: The Work of Julia Kristeva*, London: Routledge.

Clover, C. (1993) 'White Noise', *Sight and Sound* 3 (5): 6–9.

Featherstone, M. (1991) 'City Cultures and Postmodern Lifestyles', in *Consumer Culture and Postmodernism*, London: Sage, 95–111.

Finn, G. (1993) 'Why Are There No Great Women Postmodernists?', in V. Blundell, J. Shepherd and I. Taylor (eds) *Relocating Cultural Studies*, London: Routledge, 122–52.

Grosz, E. (1992) 'Bodies–Cities', in B. Colomina (ed.) *Sexuality and Space*, Princeton, NJ: Princeton Architectural Press, 241–53.

Grosz, E. (1995) 'Woman, *Chora*, Dwelling', in S. Watson and K. Gibson (eds) *Postmodern Cities and Spaces*, Oxford: Blackwell, 47–58.

Hoberman, J. (1991) 'Roadside Attractions', *Sight and Sound* 2 (4): 6–8.

Jarmusch, J. (1991) Interview with Peter Keogh, *Sight and Sound* 2 (4): 8–9.

Jones, K. B. (1990) 'Citizenship in a Woman-Friendly Polity', *Signs* 15: 781–812.

Keith, M. and Pile, S. (eds) (1993) *Place and the Politics of Identity*, London: Routledge.

Mahoney, E. (1994) 'City Limits: Women, Cities, Postmodernism', in S. Earnshaw (ed.) *Postmodern Surroundings*, Amsterdam: Editions Rodopi, 133–46.

Mazzoleni, D. (1990) 'The City and the Imaginary', *New Formations* 11 (Summer): 91–104.

Massey, D. (1991) 'Flexible Sexism', *Environment and Planning D: Society and Space* 9 (1): 31–57.

Morris, M. (1988) *The Pirate's Fiancée: Feminism, Reading, Postmodernism*, London: Verso.

Nicholson, L. (ed.) (1990) *Feminism/Postmodernism*, London: Routledge.

Orwell, G. (1948) *Nineteen Eighty-Four*, Harmondsworth: Penguin.

Patton, P. (1995) 'Imaginary Cities: Images of Postmodernity', in S. Watson and K. Gibson (eds) *Postmodern Cities and Spaces*, Oxford: Blackwell, 112–21.

Petro, P. (1989) *Joyless Streets: Women and Melodramatic Representation in Weimar Germany*, Princeton, NJ: Princeton University Press.

Pollock, G. (1988) 'Modernity and the Spaces of Femininity', in *Vision and Difference: Femininity, Feminism and the Histories of Art*, London: Routledge, 50–90.

Rose, G. (1993) *Feminism and Geography*, Cambridge: Polity Press.

Soja, E. (1989) *Postmodern Geographies: The Reassertion of Space in Critical Social Theory*, London: Verso.

Walkowitz, J. (1992) *City of Dreadful Delight: Narratives of Sexual Danger in Late Victorian England*, London: Virago.

Wilson, E. (1991) *The Sphinx in the City: Urban Life, the Control of Disorder and Women*, London: Virago.

Wolff, J. (1985) 'The Invisible *Flâneuse*: Women and the Literature of Modernity', *Theory, Culture and Society* 2 (3): 37–46.

Wolff, J. (1994) 'The Artist and the *Flâneur*: Robin, Rile and Gwen John in Paris', in K. Tester (ed.) *The Flâneur*, London: Routledge, 111–37.

9

MAINLY IN CITIES AND AT NIGHT
some notes on cities and film
•

Rob Lapsley

Our situation is not that of a rescuing coast; it is a leap into a drifting boat.
MARTIN HEIDEGGER[1]

I'm out of here
WAYNE in *Wayne's World*

RUINANCE

In his writings on factical life Heidegger spoke of life as ruinance, of *Dasein*'s sense that life has not turned out as it should. Ineluctably drawn into the world by its proclivities and inclinations *Dasein* feels it has been swept way; involvement is experienced as fallenness. Always under way in a process which it neither fully comprehends nor completely controls *Dasein* perpetually seeks to come into its own only to find expropriation in every appropriation. As things go wrong and its projected retrievals misfire, *Dasein* becomes increasingly self-burdened and unsettled. Oppressed by the sense that in its self-dispersal life has been squandered, *Dasein* is *weg-sein*; being there is being gone. Consequently, *Dasein* in its thrownness has always already been put to work repairing the ruinance in a task as impossible as it is unavoidable.

The happening of the city is at once a metaphor and a site of this ruination. As metaphor the city is a process which always veers away from the form envisaged and desired by its creators, a process outrunning understanding and control, a process whose revenge upon its architects and planners undoes every dream of mastery. As site it is the space held open by *Dasein*'s care, the envisioned space which shapes and is shaped by *Dasein*'s projection of its future possibilities, the primordial space of which all others are an abstraction. Crucially it is one of the sites where *Dasein* is assigned the impossible task of putting right what can never be put right.

The Lacanian subject similarly experiences the city as inseparable from subjective dereliction and destitution. Although every subject is a knotting of the registers of the real, the symbolic and the imaginary every subject experiences this knotting as an unravelling; in search of unity it is fated to know only division. As with *Dasein* the subject's attempts to reverse this experience of ruinance produce only further alienation and deprivation. Again the city is at once the metaphor and site of this process: metaphor of an order experienced as a disordering; site of the space which, constituted by signifiers, promises but forever defers self-realisation.

Although the later Heidegger insisted there was no equivalence between *Dasein* and the subject of psychoanalysis (for which he felt only distaste) it is both legitimate and useful for present purposes to conjoin Heidegger and Lacan or more precisely to use Heideggerian formulations to highlight particular elements in Lacan's teaching. In their very different ways they have explored more fully than any other twentieth-century thinkers the fate of social beings who alone and without a rescuing coast must confront their finitude and address the uncompletable tasks that finitude assigns.

Pre-eminent among these tasks is the necessity of articulating a relationship with others. For both *Dasein* and the subject, relationships constitute primordial unresolvable problems.

For Heidegger *Dasein* is *mitsein*, being there is being with others. *Dasein* has no choice but to interpret itself in the pre-existing vocabulary of others and to organise a relationship with others on the terms rendered possible by that vocabulary but *Dasein* inevitably comes to experience this involvement as a self-distraction and self-scattering. Dependence on the 'They' for its understanding and self-understanding gives rise to a sense of inauthenticity. *Dasein* is therefore set the impossible task of achieving a relationship with others which permits authenticity.

For Lacan matters are even more complex: the subject does not relate directly to others but to the Other, the symbolic order interposed between the subject and other subjects. Born into the world of the Other's signifiers the subject has to articulate a relationship to the Other where the mode of articulation is desire, testimony to an irremediable lack in both subject and Other. Consequently the subject, impelled by that desire, finds itself destined endlessly to attempt repair of the irreparable. For Heidegger and Lacan living is a way of being in relation to others where that relationship of care and desire can never finally come right.

IMAGES AND THE CITY

Central to this work in our culture are images and representations. It has become a cliché of contemporary writing that the city is constructed as much by images and representations as by the built environment, demographic shifts and patterns of capital investment. The politics of the city is a matter not only of capitalism's inequitable distribution of satisfactions and misery but also of the struggles around such issues as gender, ethnicity, age and religion, struggles in which images and representations informing our sense of both self and reality are a crucial factor in the reproduction and contestation of existing social practices.

It is within this perspective that I wish to consider our commerce with images and representations of the city, the works created in response to what is sent. In particular I wish to explore the thesis that in seeking to organise a happier relationship to the Other, to repair the ruinance, cultural products have the unintended consequence of producing further problems. None is cost free.

The paradigmatic instance of image production is the mirror stage. Since it makes possible the argument which follows, this needs to be rehearsed in some detail. Somewhere between the ages of six and eighteen months every child responds to the real of its situation, to the impossibility of achieving control by identifying with an idealised self-image. Lacking coordination and sunk in motor incapacity it identifies with an image of unity in the hope of thereby achieving a mastery which it anticipates but never acquires; the image of its supposed real self is an unattainable ideal; its self-recognition is a misrecognition. The subject emerges through identification with a signifier. But not just any signifier will do. Born dependent upon a caring adult for the provision of warmth and nutrition, the subject seeks the security of the Other's love through identification with the signifier of what the Other wants. As such, this attempted articulation of a relation to the Other consequently involves a knotting of the real (here the impossibility of self-mastery), the symbolic (the Other as incarnated in the speech of the mother) and the imaginary (the image of the *semblable*).

Three aspects of this pivotal moment in Lacan's teaching are of importance for what follows. First, the child does not produce this image *ex nihilo*. The terms of the specular image are not self-generated – they are given by an external Other. The child's misrecognition of itself in its counterpart is confirmed by the mother. The child identifies with a signifier assigned before its birth by the Other. If children produce images it is not in terms of their own choosing.

Second, although there is a gain, since the image enables the subject to organise both space and its intersubjective relationships, the gain is at a cost. The sense of control acquired, while indispensable to the child's functioning as a subject, is at the price of being mastered by the signifier. In attempting to organise a relation to the Other, by imagining itself to be what the Other wants, the child subjects itself to an ideal with which it can never coincide; subject and ego are never one. In seeking to come into its own, the subject is alienated; in seeking unity it is divided. The self-burdening subject's attempt to overcome the real as impossible produces further impossibilities, the solution itself becomes a problem. Identification with the *semblable* is the assumption of a style and, as Derrida reminds us, style perforates as it parries.[2]

Third, although the Other is anterior to the subject (the identity of the subject is assigned before its birth), once the drama of the subject is under way there are no givens. The subject, as much an effect as the author of the representation, changes, metamorphoses, as the process of representation unfolds. Images and representations at once open up possibilities and close others off.

According to this perspective, images, in seeking to articulate a happier relationship to the Other, are attempts to negotiate the impossibilities of the real. What holds for the self-image in the mirror stage holds also for the subject within discourse and for the subject in relation to *object a*. For Lacan these impossibilities have a single source: the lack in the Other. The human infant can become a subject only through a symbolic identification in the Other. There is no Other of the Other, no other place where the subject can emerge than the historically contingent signifying chain in which it finds itself. But, as in the mirror stage, this organisation of a relation to the Other is at a cost.

As the signifier to identify the subject absolutely is lacking (in Lacan's terms the Other is barred), the subject becomes alienated, represented but not expressed by a signifier, for 'a signifier represents the subject for another signifier'.[3] At the same time the mortifying effect of language on the human infant creates a subject destined to desire accession to a being always already lost; the *parlêtre* (the speaking being) and the *manque à être* (the want to be) are born in one and the same moment.[4] In entering language the child suffers irreversible separation from something which cannot be put into words (hence its designation as the *object a*) but which functions in the imaginary as a lure or mirage promising but never delivering the fulfilment of desire.

In response, subjects tend to shy away from this lack, wishing to believe there is a rescuing coast, an Other of the Other. Rather than accept either the incomprehensible or unmasterable nature of the Other or the unrepresentable, unattainable nature of the *object a*, the subject takes flight by imaginarising both.

Like *Dasein* the subject is unauthorised and finds that lack of authorisation hard to bear. From the moment the subject assumes an identity there is an appeal, whether explicitly or implicitly, in its every enunciation for a guarantor of that identity. No such guarantor exists, for the Other is a set of signifiers, a place not a person. To evade this lack the subject imagines another subject is the Other. As always, this process of imaginarisation produces both gain and loss. By imagining the Other to be embodied by another, the subject is able to believe in the existence of an Other who will misrecognise it as it wishes to be misrecognised. But this is at the necessary cost of insecurity under the gaze of any other elected to the place of the Other.

Shyness is an everyday example of this self-burdening. For the shy person every other is the Other and therefore the judgement of others matters inordinately. A just appreciation of others' ordinariness would free the sufferer from his affliction but typically he refuses this, because to accept it would mean relinquishing belief in the possibility of authorisation by an ideal Other. Desire for an Other empowered to guarantee his over-valued narcissistic self-image overrides the wish to be free from anxiety and embarrassment.

A similar process of imaginarisation occurs with the *object a*. Always already lost it exists only as the cause of desire. Rather than acknowledge the impossibility of its attainment the subject imagines it is located in an accessible elsewhere. A beautiful illustration is offered by Nicholas Roeg's *Walkabout* (1970).

During her period in the outback, the female protagonist experiences its otherness as threatening and feels only revulsion for her saviour, the aborigine who falls in love with her. Back in the city this is forgotten and she retroactively constructs the episode as a pastoral idyll. This enables her to believe that, while lost, plenitude is not impossible but this belief is available only at the cost of impoverishing her current existence and further distancing her from her husband.

TRADITION, MODERNITY AND DESIRE

The city is situated in relation to desire, and desire always takes a historically contingent form. The terms of the subject's responses to the *manque à être*, for example, the imaginarisation of the Other and the *object a*, are given by a historically determined signifying set. Like *Dasein* the subject is given over to interpretations and self-interpretations in historically specific vocabularies. In *Walkabout*, the terms of the girl's imaginary fantasy are determined by the Other, in symbolic forms such as the myth of the lost rural paradise. And the terms of the fantasy in turn have effects giving a particular twist to her *manque à être* and intensifying her alienation from her husband.

History, then, is the history, among other things, of the responses of men and women, to the lack in the Other, their bids to bring together in more satisfactory economies the Other, their objects of desire, and their idealised self-images; a history of attempted negotiations of lack in the Other where the Other is itself the historical response of others to that self-same problem and where the Other is consequently inflected by the asymmetries of power within a particular society. Distributional struggles around resources, power and status (the concerns of traditional history) are struggles cathected and informed by desires and identifications, fantasies and ideals. To cite an example from rural rather than urban history: when the pre-Revolutionary Russian peasant aspired to ownership of all the land he farmed, he invested those acres with a capacity to improve the quality of his life far beyond the powers of the 10 per cent of land still in gentry hands. What follows is a suggested line of exploration for the developing history of desire and identification in the urban spaces of western modernity.

Marx was in error, for history *does* set man tasks that he cannot accomplish. All history is the history of the unhappy economies produced in response to the impossible; politics is the struggle to achieve happier economies. An illustration is afforded, for example, by the question of the Other's desire.

Thrown into a world where its prematurity renders it totally dependent upon another, each child has to address the question, 'What does the Other want?' ('*Che vuoi?*'),[5] a question without answer. Since the Other desires the *object a* and as the *object a* is unrepresentable, the Other's desire is destined to remain radically enigmatic. Children respond by identifying with the signifier that is imagined to inscribe the Other's desire. While enabling the child to orientate itself socially the signifier neither completely represents the subject nor fully expresses the Other's desire. Consequently, the question '*Che vuoi?*' persistently emerges in every social encounter, provoking unassuageable anxiety. Different societies have developed different strategies to dissipate that anxiety.

Traditional societies attempted to negotiate the problem by providing a socially sanctioned answer. With the Other imaginarised as God or some elevated personage it was made clear to the subject what the Other wanted and what consequently was required of him or her. Identity was socially ascribed and the embodied subject was called upon to live up to a clearly established ideal. Exchanges were ritualised, allowing the subject to suppose knowledge of the Other's desire. The gain from this arrangement was 'ontological security'.[6] As Giddens has argued, the normative components of tradition setting out how subjects should act were the object of 'deep emotional investments' because they functioned as 'mechanisms of anxiety control'.[7]

The costs of this supposed solution were twofold. Almost invariably the mode of identity assumption perpetuated and legitimated the inequitable distribution of wealth, power and status within a hierarchised society. And, second, too much was asked of the subject. No more able to coincide with the prescribed ideal than with the specular image, the subject was condemned to a sense of fallenness interpreted as sin. Failing in its imitation of Christ the despairing subject was encumbered with a sense of guilt and shame.

With the emergence of post-traditional societies an identity was ascribed only in certain areas and for certain groups. Elsewhere the subject was increasingly enjoined to be itself, its true self. While this situation was experienced by many as a happier economy (few wished to return to the old order) there was as always a price to pay, in this instance fallenness interpreted as inauthenticity. The shadow of an unrealistic ideal again fell across the speaking being, and a culture oppressed by a sense of sin and worthlessness was exchanged for one burdened by a sense of expropriation and alienation. An authenticity the subject could never know was now experienced as lost.

And there was a further catch: the subject had to be itself in a form validated by the Other. Not just any 'true self' would do. To overcome the impossibility of securing recognition from the Other the subject tended to reduce the Other to an imaginary Other. As Contardo Calligaris has observed, 'Those who do not owe their place in the world to the symbolic ties of tradition must resort to finding their place in the Other's gaze.'[8] The guarantee once sought in the eyes of the polis or the Almighty was now sought in the gaze of an individual subject.

ROMANCE AND NARCISSISM

Typically, the Other called upon to authenticate the subject's narcissistic misrecognition was the subject's romantic love object. From time to time, remarked Nietzsche, 'there is magic'.[9] Romance, however, entails unintended consequences and unanticipated costs, because its idealisations burden each of the partners with the impossible task of embodying the other's missing complement where no such complement exists. There is always the problem of finding an other apparently worthy of being cast in the role of the Other, not just any other will do. Despite idealisation the personality of the imaginary Other does not count for nothing. And there is the further problem of being what the Other wants, an impossible task whose impossibility is never completely hidden by either the feminine masquerade or the comedy of virility.

In late modernity the enterprise of romance has become even more problematic. If lions could talk, wrote Wittgenstein, we could not understand them;[10] where there is no shared form of life what passes for communication breaks down. In the absence of the crucial signifier which would render all clear between them, men and women were always as lions to each other, but where there was tradition and shared meaning this absence was occulted. In modernity the loss of traditional sources of recognition and human warmth led to greater importance being attached to intimate sexual relationships while at the same time the centrifugal forces of capitalism worked to dissolve every intimate social bond. With the communal in growing deficit it proved increasingly difficult to hide irreconcilable differences at the same time as increasing importance came to be attached to sexual relationships. Social developments promoted investment in intimate relationships while simultaneously eroding what might ground such relationships.

To escape the vicissitudes of relationships subjects have tended to retreat into narcissism. But narcissism, like every form of identification, is an identification with a signifier, and the difficulty for the subject in late modernity is in determining the identity of that signifier. As forms of life have multiplied and individuals' perspectives become increasingly incommensurable, subjects have experienced heightened anxiety as to what the Other wants. Never in history has there been such uncertainty within a particular society; never has the Other been more illegible.

At the mercy of contingency, subjects have sought security from disconfirmation in wildly over-valued self-conceptions. By believing itself to be something rather special the subject could feel the Other's approval guaranteed but again the solution introduced new problems. As before the subject was burdened with an ideal to which it could never correspond, but with a further turn of the screw: in a world where it was increasingly apparent that every encounter is a missed encounter, subjects holding to the view 'either I am the ideal of the Other or I am nothing' often felt they were nothing. Anxiety is exchanged for depression, and oscillation between these states becomes the characteristic pathology of late modernity.

CITIES: REAL, SYMBOLIC, IMAGINARY

As the particular representative of modernity the city evokes and calls into play all these structures and processes of the subject with special intensity. The city, then, can be analysed as existing in the three Lacanian registers of the real, the symbolic and the imaginary – and subjective experience of cities can be considered as singular knottings of these registers.

As real, the city is the city impossible of achievement. Although many people plainly prefer life in the city to any available alternative, for all inhabitants there

is a non-coincidence between the actually existing city and their ideal, and consequently a sense of non-belonging. In the order of the real, therefore, the city is
the impossibility of home; although there was no home prior to the city a sense
of homelessness is engendered by the city.

Impossibilities set the subject to work. The real of the city impels the subject to
seek a means of repairing the ruinance. Like the child in the mirror stage the
subject undertakes this work in and on the terms set by the Other. As symbolic,
the city *is* this Other, the space opened up by the signifiers which precede the
subject and of which the subject in an effect; more precisely, it is the space opened
up by the incompleteness of the Other. Among the consequences of this lack in
the Other, three particularly trouble the inhabitants of the late modern city.

The first is heterogeneity. The Other as incorporated into the subject produces
an irreducible, ineradicable otherness. Neither the city nor its subjects can achieve
self-integration – each is inhabited by alterity, and so, as Kristeva states, foreign-
ness is within us.[11] This sense of an uncanny unassimilable otherness is accentuated
by the heterotopia of the modern city. The desire of man is the desire of the Other
in every setting but the modern urban subject, constituted at a multiplicity of sites
and pulled in opposing directions by the demands of disparate roles, experiences
a heightened sense of self-dispersal.

In the pre-credit sequence of *Mean Streets* (1973) Charlie reflects that one makes
up for one's sins in the streets, not just in church. He rises in the middle of the
night to contemplate in the mirror the image of self-unity which this impulse implies
but as he returns to bed the jump cuts are set to the rhythm of the Ronettes' 'Be
My Baby'. Already his bid to gather his being into a unity is overtaken by the
other of his resolution, the seductive physical immediacy of the music conflicting
with his spiritual aspirations. Purity of heart, Kierkegaard tells us, is willing one
thing;[12] but this is impossible outside the imaginary realm of narcissism and idealisation, and that impossibility is accentuated in the modern city. Charlie is torn
between his allegiances to his lover, his friend, his Mafiosi connections. His bid
for self-integration leads to tragedy for himself and others.

The second is the ineliminable presence of the gaze. 'To whom', asks Susan Buck-
Morss, 'do the streets belong?'.[13] The answer is the Other. To be a subject is to
be trapped within a field of visibility, subjected to the pre-existing gaze of the
Other. If the space of the city is defined by the gaze of the Other, in late modernity
that gaze increasingly provokes anxiety. Tradition 'protects against contingency';[14]
in its absence subjects are unprotected from the enigma of the Other's desire.
Inscribed in the eyes of the ubiquitous stranger that desire is increasingly indecipherable. Although there are institutions (such as families, schools, workplaces)
where the desire of the Other is made as plain as possible, contexts and settings
where it remains troublingly opaque are multiplying, requiring the subject to make
it up as he or she goes along.

While some welcome this as a liberating opportunity for experiment and self-
invention others are more impressed by the problematic nature of developing a
self-narrative. The terms of self-authorship are no more freely chosen than those

of the specular image; there is no auto-foundation. In the contemporary city the difficulty for the disembedded subject is to determine the enigmatic Other's forms of reference. In the theatre of late modernity the subject is called upon to perform where the demands of multiple audiences are uncertain and to speak where, although there are no pre-given lines, only certain lines will succeed. As the Woody Allen character, courting the eponymous Annie Hall (and worrying that he sounds like FM radio), is only too aware and as Travis Bickle, taking Betsy to a porno movie on their first date, is tragically unaware, not just any signifier will do.

The third unsettling instance introduced by the symbolic is desire: the city is the Other's broken promise. Desire as the name given to our relationship to the Other is a corollary of all modalities of subjectivity but the seductive glamour of the modern city produces an acute form of the *manque à être* in many of its inhabitants. Cities induce a particular mode of the melancholy which is a fundamental mode of *Dasein*'s attunement. As Elizabeth Wilson notes, this 'sorrowful engagement with cities'[15] seems to arise partly from the enormous unfulfilled promise of the urban spectacle, the consumption, the lure of pleasure and joy which somehow seem destined to be always disappointed.

As imaginary, the city is the images and representations invoked to fend off the lack introduced into the real by the symbolic. It is the bid to make whole, to suture the wound of urban subjectivity. Like the specular image the imaginary city is the attempt to overcome alterity by means of alterity, the attempt to complete what forever resists completion. As response to the unattained in what is attained the imaginary city is the creation of a further unattainable ideal.

Like *Dasein* the subject is the effect of an ungraspable process in which withdrawal accompanies every disclosure. It inhabits a non-Euclidean topological space structured by an Other which cannot be encompassed and oriented by an object which cannot be represented. As the subject is at once included and excluded from the incomprehensible Other of the signifying chain and as the *object a* is an extimate object, both internal and external to the subject, neither the subject nor the Other nor the object cause of desire can appear in specular space except in imaginary form. Completion through representation is impossible, hence the subject who craves translation from topological to specular space, like the infant before its mirror image, finds that imagined self-unity produces an excess which undoes that very unity. The following section outlines some of the strategies of idealisation and imaginarisation by which the subject seeks to master the unmasterable. In the subsequent section I will consider the costs associated with the remainder that is produced by the specularisations of urban space.

STRATEGIES TO PARRY THE REAL

Every city is singular, the happening of difference. If all subjects parry the real with the imaginary, each subject does so differently. In the modern city, for example, all three supposed remedies for the lack in the Other discussed earlier – religion,

9.1   The city affirmed: Woody Allen's *Manhattan*
© United Artists, 1979. Courtesy of BFI Stills, Posters and Designs.

romance and narcissism – coexist. While many subjects hold to the certainties of faith and religion others find solace in the romantic illusions which still saturate much of popular culture; narcissism for its part is almost everywhere.

What follows is an attempt to assess some of the strategies for coping with ruinance in the city. Common to all of them is an attempt to short-circuit the ineluctable relation to the barred Other through a fantasy of completion. Each represents a bid to evade the real of the city, by creating a zone immune from ruinance. In all instances this involves idealisation and a concomitant belief in something which does not exist. The city itself is rarely the object of idealisation. Rather surprisingly, given the preference of many people for the freedoms, excitements and energies of urban existence, filmic affirmations of city life such as Woody Allen's *Manhattan* (1979) are infrequent (Plate 9.1). Overwhelmingly, fictional representations of the city have been hostile. From the London of Griffiths' *Broken Blossoms* (1919) to the New York of *Seven* (1995) the modern city has been presented as inimical to human happiness. Instead of idealising the city the predominant strategy has been to conjure into existence an elsewhere free of lack and ruinance.

The subject and an elsewhere are equiprimordial. To be a subject is to be at once separated from unity, harmony and something of life, and untiringly to seek a place where matters are otherwise. Inhabitants of the medieval city built a place apart, a sanctuary within the cathedral 'in which the children of God were made safe from the street'.[16] Confronted by the horrific urban environments of early capitalism, subjects such as the protagonist of *Walkabout* retroactively constructed the rural as a lost inheritance where existence went with the grain of being. Lovers responding to the perceived coldness and impersonality of modern society attempted to set the city at nought in a localised space where the Other was reduced to the beloved, imagined embodiment of the *object a*.

Cinema and television have continued this strategy of creating a space set over against the city. Most obviously, they have done so in their celebration of the opportunities for self-fulfilment afforded by mythical sites from the past or future in such genres as the western, science fiction and costume drama; and they have affirmed the power of romance to create a garden in the city. Cinema and television have structured a place apart where it seems possible to take a distance from subjectivity and ruinance.

NON-REVERSIBILITY

Rather less obviously, artistic representations could be said to create a relationship of non-reversibility where the subject can imagine him or herself transcended. In dialogue with the Other the subject encounters lack both within and without, typically in three moments. In the first the subject addresses a demand to the Other whom it imagines can meet that demand. When the Other fails to meet the demand in full, the lack in the Other gapes and the subject encounters the Other as barred. Then in the final moment of this dialectic there is a reversal: the subject becomes the addressee of the demand implicit in the Other's enunciation and discovers his or her own inability to meet the Other's demand. A lack emerges in the subject and hence an equivalence between subject and Other. The identity of desiring subject and desiring Other is dialectically discovered: both are lacking, both are barred, both default.[17]

In strong contrast to this dialectic, works of art rarely confront the subject with any such danger. Instead they tend to offer the security of a non-reversible situation where the lack does not emerge. The technological form of film and television does not preclude exposure to the desire of the Other – the direct appeal of anguished parents to their child's abductor is only the most obvious example – but in general they afford the spectator a space made safe from the real of the Other's desire. In the face of difficulties *Dasein* tends to bolt and lock itself away to escape the insecurity born of all those processes – global, local and individual – outside its control; subjects nightly seek refuge in the locus of control that is non-reversibility. The further modern culture has moved away from the securities and certainties of traditional society, the more problematic the question of the Other's desire has become and the more attractive the sanctuary afforded by non-reversibility appears.

There is an evident parallel here with a certain strain in popular music. The paradox of such music is that whatever the ostensibly negative content of the lyrics – whether banal romantic loss or fashionably apocalyptic visions of social break-down and ecological catastrophe – the tone is upbeat; lyrics and music are at odds. Within the collage-like structure of such works, lyrics acknowledge the pain of existence and its many causes while the music is celebratory and affirmative. In the same moment that the lyrics announce 'It's all over', the music proclaims 'It's all right'. What is affirmed in such works is not the disaster, the loss or the attendant suffering, but the capacity of the singer(s) and, by a process of identification, the listener to take 'it'. The music suggests that the listener occupies an invulnerable position where he or she can set the world (here a metaphor for the real) at defiance. Transported by the power and immediacy of the music, the subject feels it has found a place apart from the facticity and finitude of subjectivity, from the impasses of desire.

Frequently this apparent withdrawal from the toils of subjectivity is the organising principle of pop videos where shots of various problematic aspects of social existence are alternated with shots of the singer(s) in another space enjoying an apparently self-sufficient relationship with their instruments and the music. Yes, there is suffering, but there is also an elsewhere in which it no longer matters.

SUBJECTIVITY AND TRANSCENDENCE

Paradoxically only angels and gods are satisfied by transcendence, those marked by desire demand something more. Ratings, sales, and takings all attest that songs, films, and television programmes affording a position of quasi-divine security are not equally popular; the majority fail with the majority of the audience. Such failures emphasise the absence of any fallback position from the Other; there is no Other of the Other.

Despite his or her identification with an imaginary transcendent being, the subject remains finite and continues to desire and it is only in so far as its desires are taken up that a work is of interest or significance. It is this which explains the well-known paradox of much popular music, namely that while the lyrics are often inaudible the music is impoverished by their absence. Relationship is primordial. It is only in a continuing relationship to the words or, where the lyrics are inaudible, to what the words betoken, namely the impasses of desire, that taking a distance from the fate of subjectivity holds attractions. When such songs 'work' it is through a process of dispersed identification in which the listener is at once the desiring subject of the lyrics and the imaginary Other located outside entrammelled subjectivity. For the duration of the story, identifications reduce the world to a specular space viewed from a position of security apparently outside the Other's field of visibility.

Similarly, in film and television the position of transcendence satisfies only in conjunction with the dispersed identifications of fantasy. As every subject fantasises differently there can be no synoptic account of the fantasies of later modernity. But certain modalities are particularly prominent in the imaginarisation of the city. In many instances the subject responds to the real of the city, the impossibility of home, by envisioning a place apart where self-division and self-dispersal can be avoided, where it can be its real self. This creation of a zone free from ruinance, a place where the subject can come into its own, has typically been accomplished in one of two ways.

Either a sanctuary is imagined within the city or, as in the previous example of pop music, the existence of an elsewhere outside the city is suggested. In the first the city is reduced to a specular space and the subject identifies with an idealised figure inhabiting that space. For example, the subject imagines that it has given the Other the slip by identifying with a self-certifying figure who has no need of the Other's recognition. Alternatively, it can identify with a figure representing the lost complement of the barred Other, thereby in the imaginary completing both the Other and itself. In the second, while space is specularised, it is the beyond of specular space which, although necessarily unrepresented, is imagined to be the site of a possible home-coming. Since transcendence in itself is unsatisfying, the subject continues to identify with figures within specular space but at the same time imagines there is an Other of the Other. Withdrawal is possible to a place where integrity does not fall to ruinance, where it is possible to be a subject unentrammelled by subjectivity. Consequently, the subject can at once continue to desire according to the fantasy scenarios of specular space and to transcend the ensnarling impasses of desire.

These two modalities offer possible strategies for negotiating the three related consequences of the symbolic order and lack in the Other cited earlier: the heterogeneity of the subject, the alterity of the Other and the ineluctability of desire.

THE HETEROGENEITY OF THE SUBJECT

Fantasies responding to desire as the desire of the Other seek to introduce homogeneity where there is heterogeneity principally through two strategies. In the first, an operation *within* specular space, viewers identify with idealised figures possessing a consistency unknown to any speaking being; hence the attraction of stars and autonomous protagonists. Subjects, dispersed and divided within the signifying chain, identify with stars who, always the same whatever the role, seem to enjoy the self-unity all subjects crave. 'Style', wrote Lacan, 'is the person to whom you are speaking',[18] but these figures appear self-determining. In a runaway world, where subjects increasingly experience their fate as governed by forces outside their control, spectators take flight into Promethean identifications with powerful, apparently self-sufficient figures who are the masters of their own destiny.

Alternatively, in the absence of an idealised figure as point of identification, works of art through the reduction of the world to a specular space can still afford the subject the illusion of completion. In such works it seems that reality has been encompassed and concomitantly it seems that the subject of this unifying vision has itself been unified, as though in seeing the world whole the spectator has been made whole. This experience is often the hallmark of works of high art rather than popular culture. In the name of realism high art tends to eschew idealisation in the form of heroic characters, easy solutions, and happy endings. Rather, in enunciating hard truths they appear to confront the spectator with the real, exactly the claim often advanced for their supposed superiority.

In practice, our experience of high art is one not of subjective destitution but of release and security. Such works usually insulate spectators from the real as effectively as idealisations in more popular forms, the main difference being that in popular forms the idealised point of identification is usually a figure visible within the text while in high art it tends to be the implied creator or the spectator.

Typically, such works claim to have encapsulated the human condition; not, of course, to have said everything but to have said everything that matters, the essence revealed, the truth of the human condition disclosed. By implication, as the world has been totalised, the subject is able to believe that he views it from an elsewhere – that in comprehending everything he has transcended everything. The spectator can believe himself safely distanced in a place apart, an Other of the Other, an elsewhere in which subjectivity is no longer snagged on desire.

But this is delusion, because it is only from the perspective of the subject whom the spectator *has become* that it all appears to make sense. Just as romantic love reduces the Other to the love object, the only one that matters, so textual exchanges with high art seem to reduce human existence to a single totalisable truth, the only one that matters. In art, as in romantic love, the understanding produced by a temporally contingent perspective appears to be the basis for a fixed, once-and-for-all relationship to the Other. But in both the recognition of truth is an effect of misrecognition: in art movies the sense of transcendence is the product of immanent involvement, and the sense of autonomy the result of heteronomous engagements.

When the text 'takes' (whether popular or high), when the subject's identifications and desires are reconstituted, what matters changes. The optic of the narrative becomes the perspective of the subject; the text's concerns become the subject's; vehicled by a centred narrative the decentred subject itself becomes centred.

If, therefore, *Walkabout* appears to be a summation of critical aspects of the human condition, if the well-known lines of Housman quoted over the end-credits[19] seem to state an important truth, this is a function of the identifications produced in exchange with the text. It is only in the foreshortened perspective afforded by dispersed identifications with – most obviously – the young woman and the imaginary Other who is the compassionate, transcendent witness of her suffering that the text seems to enunciate all that needs to be enunciated.

THE ALTERITY OF THE OTHER

A similar twofold strategy operates to mask the alterity of the Other encountered in the Other's gaze. First, through identification with a star or protagonist in specular space the viewer can imagine that he has captured and tamed the Other's gaze by being what the Other wants. Identified with a performer who does not stumble, who says exactly what he or she means, who succeeds in being at one with the Other, the spectator can achieve in the imaginary the self-validation available nowhere else. Alternatively, the subject can elude the question inscribed in the Other's gaze by identifying with the implied vantage point of the Other. The spectacle of the world ceases to be a trap where the *parlêtre* is subjected to the Other's gaze and becomes instead an entertainment played out for the spectator.

More frequently, the subject is dispersed between these two identifications. To be a subject in modernity is to be called upon to be one's true self and to perform to the satisfaction of the Other. Always already in relation to an inconsistent Other the subject has to mount a performance in accord with both the Other and his own unique desire. It is an impossible trick to pull off without imaginarisation, hence the tendency to identify with successful performers.

From the perspective of the spectator the idealised performer seems to have succeeded in at once achieving authenticity and, without self-compromise, securing the Other's recognition. But this appearance of authentic self-expression depends upon the spectator's short-circuiting the lack in the Other by identifying with both the idealised performer and the imaginary Other of judgement. The performer's imputed authenticity is an effect of the signification, and the self supposedly expressed is a retroactive construction, at once the cause and effect of the spectator's identifications. The performer no more coincides with the staged ideal than the subject with the Other. 'All men', Cary Grant is reputed to have said, 'would like to be Cary Grant. I would like to be Cary Grant.' However, if Archibald Leach never achieved the consistency of being that was Cary Grant, from the vantage point of the imaginary Other of their identification spectators can believe they succeeded where Leach failed. The problem of the Other's gaze is again resolved by becoming in the imaginary both subject and Other.

THE INELUCTABILITY OF DESIRE

Fantasies to overcome the impasses of desire or, more precisely, the absence of the *object a*, have already been mentioned in several contexts so require little elaboration here. Generally they acknowledge the existence of ruinance – it's a jungle out there – while positing the existence of an idealised elsewhere in which life does not go astray: a past or future home outside the city, an island of romantic rapport, a cell of good living or a personal integrity uncorrupted by the chaos of the streets. The constant in all these fantasies is plenitude. Whatever the overt content, fantasy, in apparently overcoming the heterogeneity of subject and object, always somewhere secretes the ideal of plenitude. The topological space of the barred subject, the space structured by an uncontrollable Other and oriented by the unrepresentable *object a*, is presented as a specular space where each is figured and where their relationship becomes one of harmony, division reconciliation and knotting unity.

IRONY

The problem for the late modern spectator is believing in the fantasies available. While subjects passionately long to believe in whatever will mask the insupportable real, repeated disappointment has undermined their faith in the alibis of fantasy. Much as they would like to be taken in, many no longer are; hence the strategy of ironic distancing.

Following Baudrillard and then Jameson it has become conventional to emphasise proximity and the collapse of distance as a feature of contemporary society.[20] In many circumstances the contrary is the case. *Dasein*'s tendency to take its distance and thereby imagine itself superior to others remains a salient feature of our culture. In the form of ironic distancing this tendency is evident in both the production and the reception of contemporary cultural forms.

For example, the introductions of talk-show hosts, quizmasters and disc-jockeys often constitute a subject: 'I know this is artificial and strained, a simulation of the real thing, and I'm not fooled'; 'This is not the real me just an act.' Similarly, from a remarkably young age television spectators are mocking and disdainful of even their favourite programmes: 'I know this is hopelessly tenth-rate and I've seen all this a thousand times.' Such attitudes imply that the watching and enjoying self is not the real self ('I'm better than that').

This strategy for negotiating the real appears different from earlier examples in that here the subject seems free of illusion. But the ironist's claim to be lucidly confronting the real of a disenchanted world is mere pretence since he is equally committed to belief in an illusory elsewhere. In telling himself 'This is not "the real me"' he cleaves to the fantasy that in some undefined elsewhere fullness of being is possible. As no such place exists the ironist's true self is actually the performer 'faking it' or the spectator uneasily enjoying junk TV. As Lacan says, '*les non-dupes errent*'[21] – those who imagine themselves undeceived are. The ironist who imagines himself a sovereign being disengaged from and indifferent to the game continues to play, watch and desire. The city has us before we have it.

As in the mirror stage the subject's belief that it can come into its own is at a cost, every attempted gain incurs its own loss. It is to the costs of this and the other strategies sketched above that I now turn.

COSTS

In the San Franciso of Hitchcock's *Vertigo* (1958; Plate 9.2) the lack in the Other is a terrifying abyss. Originally figured in the credit sequence as the gaze of a woman, the question 'What does the Other want?' opens on to nothing. This gaze and the question it inscribes is then linked metonymically in the subsequent rooftop chase to the void over which the protagonist Scottie is suspended. To negotiate the abyss Scottie develops a phobia, an organisation of space which purchases a zone of immunity at the cost of creating a zone threatening annihilation. His vertigo

9.2   The San Francisco of Hitchcock's *Vertigo* (1958)
© The Alfred Hitchcock Trust. Courtesy of BFI Stills, Posters and Designs.

makes areas safe while rendering others terrifying; self-protection is self-burdening. Then, in a 'clinically' well-documented move, Scottie replaces the phobic object with a fetish object. He elevates Madeleine to the status of *la femme*, a fascinating figure occulting everything other. His self-burdening, which always had implications for others, now very directly burdens others. Madeleine is called upon to realise an impossible and ultimately lethal ideal, while the devoted Midge is treated with the hurtful disregard that romantics reserve for those outside their fantasy space.

Every subject is assigned certain tasks by the lack in the Other but in seeking and failing to accomplish their ends subjects burden not only themselves but also others. As there is no place apart, as relationship is primordial, a subject's response to the real as impossible necessarily affects the others with whom he or she is inextricably entangled. And as subjects relate to the barred Other and the *object a* first and others second, those others get hurt. History is consequently the history of those with power seeking to solve their problems with the real at the expense of themselves and, even more, those they oppress. In the west the conspicuous example is Christendom which provided 'answers' for millions of people but in many instances at a terrible cost in terms of self-oppression and at an even more terrible cost for the non-Christians it persecuted.

Typically, the parrying of the real involves an idealisation whose other face is violence, since any idealisation makes demands on self and others that cannot be met. A telling example is offered by Victorian sexuality. In response to the impasses of desire and the impossibility of a sexual rapport, men took refuge in an ideal, sex became sacramental, and women the incarnation of moral purity. But in seeking to articulate a relationship to the Other by fantasising a sexuality which would make them whole, men burdened both themselves and others. As Freud's clinic attested, they burdened themselves in that they had to shun sexual contact with the women they deified (lest they discover the troubling alterity of those women's desires) and instead sought lesser satisfactions with others whom they considered debased. More importantly, they burdened women by denying them any articulation of their desire except within a male-organised fantasy where they were called upon to combine two contradictory and equally alienating roles.

The relationship of city and image is similarly one of ideals and violence. Like the mirror stage the city is a bid for mastery, an attempt to create a space where the subject can come into its own, an instance of the *maître moi-même*. As such, whatever the opportunities and benefits the city produces, collisions with the impossible are inevitable. No city, no life goes according to plan; the city, like the subject, is always engaged in the double movement of setting limits and then surpassing them. All attempts at mastery and unity produce an unmasterable otherness; the supposed solution becomes a problem. Cities are created in the image, not the self-image, of man.

Consider, for example, the idealisations inherent in the strategies for circumventing the lack in the Other discussed in the preceding section. In the first the subject identified with an idealised figure within specular space. Romance and the action movie provide examples of the salient strategies. In romance the lack in the Other is acknowledged but in a form that could be made good; harmony, reconciliation, and completion are held to be possible. There is lack, something is missing in the life of the protagonist, but all can be remedied by discovering the lost complement. The costs of this blissful illusion are borne by both the subject and its others.

Romantics impoverish their own lives by disparaging what *is* in the name of what can never be, by their indictment of actually existing relationships, and by their scorn for human warmth and solidarity. Worse, they inflict violence upon those they idealise. Male romantics, for example, to evade the lack in the Other demand that women dissimulate the lack. To mask the absence of a sexual relation women are called on to embody the *object a*; to mask the absence of guarantees in the Other, women are called upon to embody the Other so that they are thus assigned the impossible task of incarnating what does not exist. Failure is thus doubly inevitable, and consequently women become the objects of male hatred. Aggression aimed at the lack in the Other becomes displaced on to those who have failed to redeem that lack, hence the textual pattern described by Laura

Mulvey in which women are alternately idealised and punished.[22] Women are attacked, not because of some intrinsic feminine property, but because in their alterity and difference they fail to correspond to the idealisations of heterosexual males.

In the action movie, on the other hand, the lack in the Other is denied in certain special cases. While the existence of beings lacking consistency and authorisation is acknowledged, others, most notably self-sufficient heroes, are held to possess both. In these films, unlike romance, the achievement of sexual rapport takes second place to the celebration of male narcissism. Women are marginalised as victims, trophies, or badges signifying the hero's heterosexuality and desirability. Instead of romance violence is the thing. As an idealised self-image places the subject under threat of disconfirmation the corollary of narcissism is aggression. For the narcissist 'It's him or me', the scenario action films endlessly rehearse. Identification with these self-possessed masterful masculine ideals provides a refuge from the Other of the city but at a cost to both the subject and its others.

The hapless narcissist not only burdens himself with a depressingly unrealisable ideal but also handicaps himself in future relationships in that his bid for security proves a form of self-endangering. At risk in exchanges outside their control subjects seek security in increasingly overvalued self-images. Assuming that being something rather special will guarantee recognition, subjects identify with stars and heroes only to find, if misrecognitions are not shared by others, that this diminishes their chances of confirmation. Condemned by subjectivity to seek validation, the urban narcissist is as much a seller of his idealised self-image as of his labour time.

However, in the bear market of late modern intersubjectivity sales are hard to come by; others cast in the role of the Other are increasingly reluctant to buy into the subject's misrecognitions. Often potential purchasers have themselves taken refuge in absurdly over-valued self-images. Consequently, they can buy the seller's conception of himself as somehow rather special only at the cost of relinquishing their own very similar self-appraisal. For each of the rivals it is, as always in narcissism, 'him or me', and each is intent upon victory for the self. But in this late modern reprise of the master/slave dialectic victory is impossible. If the Other acknowledges the subject's supposed superiority he owns to his relative inferiority and ceases to be the Other; if he withholds recognition he may appear to be the Other but the subject is left destitute. Where members of traditional societies aimed for salvation and justification through integration in a pre-existing order, the subjects of modernity seek to resolve their problem through mastery and control – only to find that the Other can be neither mastered nor controlled.

If the narcissist's apparently self-serving images are self-injuring they are even more damaging to others caught in the crossfire between the narcissist and the barred Other. In action movie car-chases and shoot-outs the hero, authorised by the overriding importance of his pursuit of the demonised Other, casually visits havoc on innocent bystanders, providing not only an inspiration to male narcissists but a near-perfect metaphor for their progress. Narcissists identifying with the ideals of masculinity informed by these images experience the alterity of others as insupportable: 'The Other as such is more and more perceived as a potential threat, as encroaching upon the space of my self-identity.'[23] Intent upon reducing otherness to the same, narcissists view others who refuse to play their assigned roles as simply in the way.

The alternative mode of overcoming the lack in the Other, the imagining of a space outside the Other of the city, can be dealt with more briefly because in its effects it is little more than a variation on the narcissism just discussed. In this modality the lack in the Other is recognised as irreparable but there is an elsewhere, a transcendent position, an Other of the Other. It is best represented by those spectators of art movies who believe they have learned from the work and become the better for it. Believing they have achieved a privileged insight into the nature of things, when they have only fleetingly achieved a happier relation to the Other, they alienate others unimpressed by their implicit claim to superiority. Among narcissistic strategies for articulating a relationship to the Other the assumption of a self-image of superiority unshared by others is particularly disabling.

To fend off the real as impossible, subjects take refuge in idealisation – all that is other to the ideal is consigned to the domain of 'I want to know nothing about it'. In this perspective art's images and representations are an act of forgetting but while the subject may wish to forget the real it never forgets him. In each instance of the flight into idealisation, from the mirror stage onwards, something is left over, an excess which like urban sprawl attacks the limits established by the ideal.

Romance is precisely the refusal of the alterity of the Other's desire; narcissism the denial of otherness within and without the self. Both strategies, whether in conjunction with cinematic images or independent of them, have contributed to the growth of what Žižek has termed pathological narcissism 'within which the Other as such – the real desiring Other – is experienced as a traumatic disturbance'.[24] In romance and narcissism, as in the mirror stage, the attempt to achieve a once-and-for-all solution to the alterity of the Other exacerbates the sense of alterity. Each instance produces an unassimilable remainder and a subject less tolerant of that remainder. Desire, the third in every relation of subject and Other, the ruin of romantic and narcissistic fantasies of unity, becomes an object of horror.

The costs of romantic and narcissistic ideals are therefore distributed across billions of failed and failing relationships in late modern society. Capitalism corrodes relationships; the mobility of capital destroys communities; the savagery of competition sets individual against individual. In response subjects take refuge in fantasies, further divide themselves from their fellows, and continue the destructive work of capitalism by other means. Nor is the erosion of solidarity confined to intimate sexual relationships. In our culture it often seems that wherever two or more are gathered together it is in the name of narcissism and at the expense of an absent other who is doomed to be 'slagged off' for his or her difference.

In traditional societies relationships 'worked' in so far as they lasted. They did not work in so far as the oppression and exploitation involved far out-weighed their benefits. Today, though with greater fluidity there is less oppression and exploitation in intimate relationships, more rides on them. To escape the concomitant anxiety-provoking uncertainties subjects (as in the mirror stage) seek to take control through identification with an ideal. But, as we have seem, self-idealisation, far from being a means of escaping risk, is a form of self-endangering. As the congruence between subjects' misrecognitions diminishes, the clash of desires becomes more discordant. Relationships become harder to establish, and where they do exist, they tend to be impassioned, conflictual, and short-lived. Once the space of the city was defined by the lack within the subject, interpreted as his sinfulness; later it was defined by the lack in the sexual partner, typically the real of her desire. Today it is increasingly defined by the lack of a partner.

There is no escape from either ideals or illusions. What both *Dasein* and the subject are called upon to become issues from someone or something other than themselves, and the summons from this Other always involves an ideal. It is impossible to raise a child without praise and censure and the ideals implicit in both. Ruinance is consequently inevitable. If there can be no subject without ideals equally there can be no subject which successfully coincides with its ideals and no subject unburdened by this failure. If Bob Dylan's celebration of love as existing without ideals or violence in 'Love Minus Zero/No Limits' seems attractive it is precisely because nobody has managed this trick and certainly no lover.

Similarly, the existence of man and illusion are coextensive. Without illusions the subject could not operate. The mirror stage is an indispensable moment in the child's orientation within social space. Illusions are equally indispensable to society. As Žižek has emphasised in his repeated references to the tale of the emperor's new clothes,[25] the intersubjective network dissolves if illusions are exposed. Such too is the testimony of history which suggests that man is no more capable of living free of illusion than he is of holding his breath. Religion and other cultural forms shot through with wish-fulfilment are omnipresent in human societies; recourse is always had to the fantasy of a rescuing coast.

Consequently there can be no question of theory seeking to free subjects from all ideals and illusion. Rather the business of theory should be the evaluation of the effects of particular ideals and illusions and, where possible, their displacement by less damaging configurations. The aim must be economies which inflict less cruelty upon ourselves and others.

Amelioration is possible. Improvements in the physical environment and public services cannot abolish the *manque à être* but they can eradicate some of its forms and, elsewhere, render it less acute. Equally, the terms in which the *manque à être* is constituted can be altered. Given the lack in the Other all language has the structure of metaphor, and metaphors can be displaced. There is no necessity for the Other of the city to be metaphorised as an enigmatic woman (Faye Dunaway in *Chinatown*, 1974) or the real of the Other's desire to be figured as the maniacal passion of a demented homicidal woman (Glenn Close in *Fatal Attraction*, 1987). Metaphors can be changed and happier economies achieved.

Yet the scope for change is more restricted than the fashionable post-structuralist exponents of self-invention allow. Human existence is driven by desire and identification, both of which are involuntary. The specular image captivates; the object of desire lures; neither is freely chosen. In the topological space organised by the Other and orientated by the *object a*, the subject is as much the effect of the event as its author. Subjectivity is more bounded by impossibilities than open to possibilities. Amidst the ruins of the city the subject, always already begun, always already gone astray, must find its way in the labyrinth its way creates.

NOTES

1 Martin Heidegger, cited in David Farrell Krell, *Daimon Life, Heidegger and Life Philosophy* (Bloomington: Indiana University Press, 1992), p. 37.

2 See Jacques Derrida, *Spurs: Nietzsche's Styles*, trans. Barbara Harlow (Venice: Flammarion, 1976), p. 35.

3 Jacques Lacan, *Ecrits: A Selection*, trans. Alan Sheridan (London: Tavistock, 1977), p. 316.

4 Jaques Lacan, *The Four Fundamental Concepts of Psychoanalysis* (Harmondsworth: Penguin, 1979), p. 29.

5 Lacan, *Ecrits*, third graph, p. 313.

6 Anthony Giddens, 'Living in a Post-Traditional Society', in Ulrich Beck, Anthony Giddens and Scott Lash, *Reflexive Modernisation* (Oxford: Polity Press, 1994), p. 65.

7 Giddens, 'Post-Traditional Society', p. 66.

8 Cortardo Calligaris, 'Memory Lane: a Vindication of Urban Life', *Critical Quarterly* 36, 4 (Winter 1994), p. 65.

9 Friedrich Nietzsche, cited in E. F. Podach, *The Madness of Nietzsche* (London and New York: Putnam, 1931), p. 234.

10 See Ludwig Wittgenstein, *Philosophical Investigations* (Oxford: Blackwell, 1972), p. 223.

11 See Julia Kristeva, *Strangers to Ourselves*, trans. Leon S. Roudiez (New York: Columbia University Press, 1991), p. 170.

12 See Soren Kierkegaard, *Purity of Heart is to Will One Thing*, trans. Douglas Steene (London: Fontana, 1948).

13 Buck-Morss, 'The *Flâneur*, the Sandwichman and the Whore: the Politics of Loitering', *New German Critique* 39 (Fall 1986), p. 114.

14 Giddens, 'Post-Traditional Society', p. 104.

15 Elizabeth Wilson, 'The Invisible *Flâneur*', in Sophie Watson and Katherine Gibson (eds) *Post-Modern Cities and Spaces* (Oxford: Blackwell, 1995), p. 73.

16 Richard Sennett, *The Conscience of the Eye: The Design and Social Life of Cities* (London: Faber, 1991), p. 11.

17 See Alain Juranville, *Lacan et la philosophie* (Paris: Presses Universitaires de France, 1984), pp. 124-6.

18 Jacques Lacan, *Ecrits* (Paris: Editions du Seuil, 1966, two vols), vol. 1, p. 9.

19 A. E. Housman, 'Into My Heart', *A Shropshire Lad* (London: Garnstone, 1976), p. 57.

20 See Jean Baudrillard, *The Ecstasy of Communication* (New York: Semiotext(e), 1988), especially p. 27; and Fredric Jameson, *Postmodernism, or the Cultural Logic of Late Capitalism* (London: Verso, 1991), especially pp. 25-31.

21 See Jacques Lacan, '*Les non-dupes errent*' (Seminar 21, unpublished).

22 See Laura Mulvey, 'Visual Pleasure and Narrative Cinema', in Antony Easthope (ed.) *Contemporary Film Theory* (London: Longman, 1993), pp. 111–24; p. 119.

23 Slavoj Žižek, *The Metastases of Enjoyment* (London: Verso, 1994), p. 216.

24 Žižek, *Metastases*, p. 7.

25 See Slavoj Žižek, *The Sublime Object of Ideology* (London: Verso, 1989), p. 198 and *Metastases*, p. 58.

10

PIECING TOGETHER WHAT REMAINS OF THE CINEMATIC CITY

•

James Hay

THINKING ABOUT THE *PLACE* OF CINEMA AND CINEMA STUDIES *(WHILE TRYING TO DUCK THE FIREWORKS OF ANOTHER CENTENNIAL CELEBRATION)*

> The city was and remains *object*, but not in the way of particular, pliable and instrumental object: such as a pencil or a sheet of paper. Its objectivity, or "objectality," might rather be closer to that of the *language* which individuals and groups receive before modifying it, or *language* (a particular language, the work of a particular society, spoken by particular groups). One could also compare this "objectality" to that of a cultural reality, such as the *written book*, instead of the old abstract object of the philosophers or the immediate and everyday object. However, one must take precautions. If I compare the city to a book, to a writing (a semiological system), I do not have the right to forget the aspect of mediation. I can separate it neither from what it contains nor from what contains it, by isolating it as a complete system.
> HENRI LEFEBVRE, *The Specificity of the City*[1]

I want to begin with this passage from Lefebvre not so much for what it has to say about the city (an issue I will address below) but for how it could just as easily refer to cinema studies' difficulty in treating cinema as a "mediated" object and, in that sense, as a site, a formation, and property/real estate. It is that conception of cinema or *cinematic city* (a "contained" sphere of practice) that I want to address before proposing an alternative.

This chapter grows out of several conference papers that I was invited to present over 1995 – a year celebrated at each of these conferences as the Centennial Anniversary of Cinema. And these were, by no means, the only conferences that hitched their wagons to the year-long commemoration, though I suspect that were someone to pinpoint all of them on a map, that person would find their concentration greatest around France. Some were relatively small and private affairs, such as a conference at Indiana University called "European Cinema/European Society – 1895–1995." Others were relatively large and private affairs, such as the annual

Society for Cinema Studies conference entitled "One Hundred Years of Cinema – Writing the Histories." Still another was, in the tradition of the Italian historical film spectacle, an epic project to be set in Rome (and organized heroically by one of Italy's most well-known film historians).

In Europe, the Centennial celebrations continue a well-established tradition of festival patronage by cities interested in promoting their image as centers of "culture" (*and* tourism). Academic conferences operate from a different cultural site, but their project is historically implicated in the festival enterprise of these European towns. It is important (particularly in the US where the boundary between film scholarship and such commercial ventures is more sharply defined) to recognize their connection not only because both are examples of how cinema and the study of its history have become dispersed – constituting separate but contiguous institutions – but because their collective interest in reproducing cinema history belies a set of conditions – constituted by and external to them – that have transformed the cultural status and location of cinema. Basically, it seems less consequential to try to figure out who decided that there should be a 100-year anniversary celebration of cinema than to understand what the celebration suggests about the current *place* of cinema and cinema studies – the subject of this chapter.

I am particularly interested in two related issues surrounding the Centennial. One concerns European cinema and its history as an object of cinema studies (and, more generally, how certain historical paradigms such as "pre-cinema" or "classical cinema" organize cinema studies). The other concerns the deeply historicist disposition of film studies, made particularly evident in 1995. I do not doubt that the one group most suspicious of attributing an *origin* to *the* cinema or of describing cinema as "invented once upon a time" would be a group of film theorists or historians. This is, after all, a group prone to citing or accepting as "common sense" Comolli's argument (1971, no less) that film historians always "originate" cinema through materials and references at hand.[2] Thus, histories of early cinema written over the last ten years are rightly fraught with qualifications and disclaimers about what constitutes a "pre-cinema." Still, anniversaries such as the 1995 Centennial, and the spirit of historical scholarship in film studies, belies (as Foucault would note) a discursive formation whose disciplining of knowledge rests upon the institutionalization and maintenance of certain pedigrees through its production of history. The ceremonial role of an anniversary, in this sense, works to maintain "cinema studies" as a social institution and, within the academic institution, as a discipline – no small feat for a "field" of theory and research that has always struggled for a *place* within more established and often more publicly accepted fields, disciplines, and departments.

Certainly *where* cinema studies is practiced, *where* it gets produced, has everything to do with how it understands and writes history – and particularly its own history. Writing history is a process of territorializing knowledge claims about the past, not only in the sense that writing history involves collecting and selecting records that matter for particular social subjects but because it involves designating and regulating how and where these records should be implemented in the present amidst a field of sites where other claims about the past are made and (more or less) defended. One cannot therefore claim that just anything counts as history; history is made to matter through institutional sites whose relation to one another is itself a matter of struggle and change.

Writing cinema history is a condition and an effect of maintaining cinema studies' institutional positions and alignments, in relation to social institutions not only outside the university (in part against charges that teaching forms of "popular culture" somehow erodes the mission of higher education) but also within the university. Cinema has always been studied from the university either with an anxiety about its relation to more established and socially sanctioned disciplines within the university or with a confidence that rests upon its relative success in some places of being affiliated with the Literary, with an aesthetic pedigree and with literary studies' preferred ways of doing aesthetic history. (And there is certainly a pedigree of film theory, from Bazin to poststructuralism, that has under-pinned Film Studies' institutional affiliation with the Literary.)

For well over the last fifteen years, the English department on my campus has offered a course entitled "Film and the British Novel." I cite this course here less because it is an example of how film gets tied to literature in English departments than because the history of film is conjoined to a literary curriculum structured around canonical periods. Industry histories are not encouraged in English depart-ments. Different Area Studies located in the Humanities institutionalize separate histories of cinema. The multiculturalist and internationalist features of any histor-ical survey of cinema (e.g. a World Cinema overview) are themselves inflected by departments and fields. In Mass Communication departments, courses in film history are embedded in the maintenance of the history of "mass communication" or (even more grandiose and vague) the history of "communication." In all of these cases, teachers and courses territorialize writings about history as part of the university's administration of the place of a student or reader in relation to fields and disciplines of (historical) knowledge.

The paradox of these affiliations is that what usually has been produced from these institutional sites is either a discrete history (i.e. an internal story of cinema) or one whose connection to other cultural forms, such as literature, is aesthetic and formalist – in other words, without a clear sense of cinema's relation to other social sites and, for this reason, of cinema study's institutional constraints and priorities. One might conclude therefore that cinema history lies across these insti-tutional sites, and indeed that is the case if one recognizes that the production of history is dispersed (struggled over) and that various historical accounts are regu-lated across institutional sites. But in no way is this the same as saying that cinema history can be anything or that all of these are partial versions of *the* history of *the* cinema.

What would be most useful are strategies for thinking about a historical dispersal (a historical geography) of "the cinematic," that is, how certain sites are (or have become) distinguished and engaged as cinematic in their relation to other sites. To study "the cinematic" would involve considering the place(s) of film practices within an environment and their relation to other ways of organizing this environment, of organizing social relations into an environment. To consider cinema and cinema studies in this way would certainly be inter-disciplinary but would not assume that working across disciplines of knowledge and intellectual traditions is easy or that, in doing so, one ever arrives at a comprehensive understanding of cinema or the field of social relations wherein cinema could be said to have had effects. To believe that such a comprehensive understanding of cinema could be reached, however inter-disciplinary, would not be very different from historians who have attempted to demonstrate that film has a discrete history.

The tendency to construct a self-contained history emerges through efforts during the 1970s to understand film as a distinct "language" or set of formal conventions – a critical strategy that was implicated then in the emergence of film studies as an academic discipline.[3] I am not, however, suggesting that all historical scholarship about film is simply interested in unities or coherent periodization. Some historical work, genre study for instance, has emphasized discontinuity and the emerging and residual interplay of film practices. I am not ignoring efforts over the last ten years to consider national cinema histories or other film histories that have taken identity and the politics of historical reconstruction as a central issue. And some historical scholarship on cinema has acknowledged that a history of cinema is bound up with other histories – histories of screen practices, histories of perception, histories of models of vision and sound, histories of technologies, histories of governing and political institutions, histories of trade. But in all of these cases, the emphasis remains upon cinema, a strategy that has the effect of making cinema seem to have a self-contained, self-perpetuating history. Collectively this historical work privileges cinema as a historical object while affirming that there really is no such thing as *a* history of *the* cinema, that it can never be a coherent and stable object of study.

What concerns me therefore is the tendency to see film practices (or other practices with which it is seen to have a historical relation) as discrete, albeit changing, unities and as a discrete set of relations producing a "cinematic subject" rather than understanding film as practiced among different social sites, always in relation to other sites, and engaged by social subjects who move among sites and whose mobility, access to, and investment in cinema conditions is conditioned by these relations among sites. To shift strategies in this way would involve not only decentering film as an object of study but also focusing instead on how film practice occurs from and through particular sites – of re-emphasizing the *site* of film practice *as* a spatial issue or problematic.

Interestingly enough, one frequent instance when the site of cinema is discussed in relation to other sites has been in studies of a "pre-history" or an emergent cinema. Here I am thinking about Tom Gunning's discussion of a "cinema of attractions" or Thomas Elsaesser's treatment of early European cinema's relation to (Dada) performance in cabarets, nickelodeons, lunar parks, and other sites of popular entertainment.[4] Some more recent work on the relation of cinema to groups immigrating to the US from Europe around the turn of the century has touched on how local cinemas intersected with global flows of people thus becoming sites of assimilation and cultural reinvention.[5]

What I find most compelling about this historical work is not its focus on film form or on film's relation to other forms (such as theater or literature) but its impulse toward a kind of spatial materialism of the site, the concrete location where film is practiced, always in relation to other sites. Unfortunately, work about a "pre-cinema" too often has been understood through a more expansive system and historical paradigm: "film classicism." Work about a "classical" cinema, or work that uses the concept as a way of organizing other film histories, has been the most inclined to produce discrete histories. The term "classicism" is most often invoked to designate both a stage in cinema's internal development and that stage's affiliation to an earlier, deeply embedded cultural aesthetic – a move that thus has the effect of casting film practices into a kind of supra-history and that offers only the loosest and most generalized considerations about the historical dispersal of the cinematic in everyday life and about the role of the cinematic in organizing social relations environmentally.

The term also became central to cinema studies during the 1970s and 1980s through film theory and criticism that concerned itself with cinema's ideological effect – its production of a cinematic/ideological Subject.[6] I would not be the first to note that film theory and criticism's use of the term "classicism" to explain cinema's ideological effect (or their having posited an ideological apparatus that remained intact and self-perpetuating throughout the "classical" era) lacked any historical or contextual features beyond their most general connection to a "dialectical" or "Historical" materialism (with a capital "H") and the sweeping historical force of a capitalist mode of production.

Although historical studies of cinema "classicism" have offered a nuanced description of the components of the cinematic apparatus, they none the less have presumed its coherent development and its coherence within any historical context and any social relations.[7] While these studies examined the relation between a mode of production and film form (and between both of these a generalized conception of the cinematic subject), there seemed no other determinations for the history of "the cinematic" outside of this dialectic. As an aesthetic and an ideological effect regulated by a mode of production, cinema perpetuated itself.

Film theory, and the deconstructive criticism that accompanied it, generalized what Metz referred to as "cinema" (a stable, coherent, and immutable set of material, non-discursive conditions) and thus devoted all attention to the aesthetic or textual features of "films" as the basis for remaking History.[8] Histories of a "classical Hollywood cinema" have examined the relation between film form and its mode of production, understanding production in more complex and historical terms but leaving intact the notion of a cinematic apparatus.[9] In so doing, both of these strands of film studies generalized historical determinations outside the cinematic apparatus. Or one could say that they perceived the issue of determination and power in classically dialectical fashion: film form informing and informed by mode of production, or, revolutionary aesthetics as a strategy for intervening in the social by transforming historical consciousness (the historical subject) that was otherwise interpellated by the cinematic apparatus. And within this dialectical relation, cinema became both an aesthetically privileged and a socially commonplace "expression" of (was thus made to stand in for) a broader structure of power that *presumably* expressed and reproduced itself elsewhere. Not to have considered the limits of the cinematic apparatus, or how cinematic expression occurred on a "terrain" of multiple sites, made it difficult to understand its relation to historical subjects except in terms of their capture within the apparatus itself.

In the 1970s, film theory and criticism saw Marxist theory and aesthetics as bringing a (or rather, the *only*) self-consciousness about the historical/contemporary role of cinema in social relations. And film theory, which called attention to how the relation between film and social subjects (spectators) was maintained by and reproduced a deeper structure, rightly eschewed film scholarship that simply described random relations between film and society or saw films as simply reflecting events in the Real world. But film criticism's appropriation of semiotics, psychoanalysis, and certain Marxist concepts became the basis for understanding issues of form, meaning, consciousness, and pleasure of film practices in terms of a dynamic *internal* to "film language" and its subjects. Therefore, as much as it may have wanted to see film through a theory of "historical materialism," it was just as prone as less elaborately theorized historical scholarship to emphasize (indeed valorize) a history of certain film *artists* and *styles*, in short to emphasize aesthetics, meaning, and ideology over the "exteriority" of film practice (or, just as important, to have done so without having considered this exteriority as a landscape or environment that situated film practice and was only partially shaped through the cinematic).[10]

More than literary criticism, film criticism understood film narrative, meaning, and ideology as regulated by a mode of production and an "apparatus" of visibility and consciousness. However, like literary criticism, from which it drew many of its key concepts and to which it was often wed within academic institutions, film criticism's focus upon the internal properties of texts or upon the relation among film texts suggested a kind of placelessness or an absolutely separate place of the literary/filmic in social relations and history. This regards not only the assumption in criticism that reading a book or watching a film occurs in (or as if each were in) a private place, but also the assumption (supported by the first) that criticism was done from no place in particular and that critical theory applied to texts from anywhere. In this sense, film criticism and deconstructive films assumed a counterpublic sphere from which to intervene in social relations without addressing the locational politics of such a sphere.

Film studies' emergence through post-structuralist literary criticism, particularly Marxist literary and film theory's shared valorization of a revolutionary and transformative aesthetic and, simultaneously, efforts to develop a Marxist critical theory of "film form" (in part as a basis for its own cultural/academic pedigree), occurred by recuperating the European avant-garde's discourse on modernity. The formulation of a "progressive" or "counter-" cinema during the 1960s and 1970s involved reconsidering montage theories of Russian constructivism, the performance strategies of Bertolt Brecht, and Walter Benjamin's "Work of Art in the Age of Mechanical Reproduction". But film criticism's reading or appropriation of European modernism had more to do with Modernist conceptions of History than with Space. Little attention was devoted to Modernist projects in architecture and urban planning – projects that might have been more difficult for the New Left to revere. For film studies, Benjamin's work on History had more to do with regular reprintings of "The Work of Art in the Age of Mechanical Reproduction" (wherein film studies found the telos of History to be cinema itself) than with his effort in "Berlin Chronicle" to understand the production of the past as a spatial practice located in the city's complex passages (*flâneurism*). The fact that the latter was not translated into English until the late 1970s only partly explains why its view of the critic's relation to a Modern (urban, industrial) public sphere in decidedly spatial terms was not central to film criticism. Just as significant in this regard is film studies' familiarity with Kracauer's history of German cinema and his post-war film theory and not his recently translated early essays, such as the one on Berlin Picture Palaces, which considers the spatial similarities between movie theaters and factories, film's social relation to theaters, and theaters' "external" relation to everyday life.[11] And finally, what film studies gleaned from French anthropology and structuralism had more to do with understanding the relation between narrative, myth, and culture as a temporal problematic (i.e. how narrative logics constrain and enable cultural continuity and change) than with their usefulness in rethinking culture in spatial terms (as Roland Barthes does when he rereads *Triste Tropiques* as an "urban semiology" or as the Situationists do in their *dérives*/"urban studies").[12]

In another respect, film studies' engagement with Modernist theory and aesthetic practice belied its ethnocentricism in a global cultural economy. By drawing upon European (avant-garde) theory and aesthetics, North American and British film studies overlooked the complex geographic implications of film theory and film practice. As Paul Willemen has noted, simple models of class domination at home, and imperialism abroad are inadequate for addressing the complexity of shifting dynamics between intra- and inter-national differences and power relations.[13] Certainly there were studies on films by Japanese, Indian, and South American directors, but even in these instances little effort was made to situate films within national cultures or to consider the methodological issues involved in analyzing other national film cultures through the intellectual traditions of European film theory. Likewise, studies on Hollywood cinema either ignored altogether its status as a national cinema or grossly simplified, through a media/cultural imperialism argument, its status in other national cultures. Some genre studies of Hollywood cinema came closer to acknowledging cinema's relation to the national, though they did so by emphasizing cinema's role in mediating cultural continuity and change and by turning the issue of a national (film) culture into a historical one.

I do not want to suggest, however, a facile dichotomy that understands Modernism as fundamentally concerned with History (and Postmodernism as being about a "forgetting" of that lesson or as having broken that spell).[14] I am not proposing that literary and film criticism have ever relied upon spatial concepts or a spatial rhetoric, only that their conception of space is an aesthetic one, that they have too often assumed an essential spatial model or particular codes (e.g. for film classicism, for European Modernist film, for Postmodernist film), and that they are concerned mostly with an internal dynamic of cinema and with theories of subject positioning that understand the place of identity formation as abstract and equally interiorized (e.g. the darkness of the movie theater, the dream-like nature of identification with screen images, the centering or decentering of a viewer's imaginary relation to screen space).[15]

So let me offer a proposal. What is necessary, I would argue, is a way of discussing film as a social practice that begins by considering how social relations are spatially organized – through sites of production and consumption – and how film is practiced from and across particular sites and always in relation to other sites. In this respect, cinema is not seen in a dichotomous relation with the social, but as dispersed within an *environment* of sites that *defines* (in spatial terms) the meanings, uses, and place of "the cinematic." The cinema is a place distinct from other sites but, in its relation to these other sites, part of the formation of a territory which it works to map. (This is different than saying that one should only be looking for relations among the sites of cinema, also different from discussing a relation between texts and contexts, and a different emphasis than the relation between texts and "reading formations".)[16]

The social subject forms spatial frames of reference about her/his world but only as a consequence of and as a basis for navigating and engaging those sites. To say that a social subject is "within" the cinematic is not therefore to focus just on the relation between spectator and screen. Instead, emphasis shifts from the formation of consciousness, identity, and ideology to how individuals or social groups have access to and move to and from the place(s) where they engage films in their everyday lives. Film's role in maintaining and modifying social relations has to do both with how it becomes part of an environment and how it enables or constrains navigation of that landscape (i.e. other sites/sights).

In this respect, it is important to consider two issues. First, the extroverted nature of the site(s) of film practice, that is, their connection to other sites and how they are intersected by spatial flows and circulation. Second, the way that particular films serve as maps within (and thus territorialize) the places where they are engaged, and conversely how the "cinematic" is *defined* by a relation among sites and flows.

I want to outline how these issues might be addressed in the cultural context of pre-Second World War Europe and, in so doing, to rethink film theory's understanding of a Modernist aesthetic and its relation to film as a social (i.e. spatial) practice. More specifically, I want to return attention to the Italian context between the two World Wars in order to rethink my own working through of issues concerning film's relation to the formation of a national–popular culture.[17]

My inspiration on the problematic of space in this particular historical context comes partly from Manfredo Tafuri, an architectural historian, whose remarkable reassessment of Modernist aesthetics and ideology in *The Sphere and the Labyrinth* underscores their contradictory relation to an environment being reshaped by urban architecture and planning.[18] Tafuri was particularly interested in the spatial administration of power, in how national projects of social engineering and management were predicated upon new spatial practices and upon changing modes of production that enabled and constrained those practices. He attempted to understand the contradictions of "revolutionary" ideology by considering the convergence and tension between new and nostalgic spatial ideals, between rhetorics of social liberation and regulation surrounding spatial practices, and between emerging and residual building practices. As a loosely coordinated series of essays about Russia, Germany, and the US, his history is less about the temporal continuity or transformation of artistic, political, or economic ideologies/movements than an effort to examine their uneven relation in and to a changing and conflicted environment (understood spatially).

I also want to draw from insights by Mikhail Bakhtin about the literary "chronotope," but only to the extent that that term opens up the possibility of discussing the text as a site informed by and informing spatial production (and thus cultural reproduction and History).[19] What interests me about his concept is not simply that *texts* are temporally and spatially organized but that this organization is produced/consumed from a context that itself is organized by and, in turn, organizes the textual production/consumption of space. Although Bakhtin devoted more attention to the conventionalization of certain spatial references in the novel than to the spatial organization of contexts where novels were produced and consumed, his argument recognizes the spatial implications about the relation between texts and contexts. In this sense, he understands the chronotope more as a historically materialist view of the literary form than through a spatial materialism of contexts of literary production and consumption.

To reconsider the emergence of a national–popular culture in Italy after the First World War would involve examining film's role in mapping the nation as territory – or "empire" – constituted of "popular" sites. This would entail examining how film emerged through a landscape of everyday life, of other sites that became "popular" by virtue of their being designated as such and being occupied by social subjects whose identities were formed from and through these sites. For this reason, it would not *privilege* film's role as a "popular" site in everyday life.

In some respects, film's role in mapping the national–popular as territory has been obliquely addressed by those of us who have written about Italian cinema during the 1920s and 1930s. Feature-length films, documentaries, and newsreels all developed conventions for mythologizing urban and rural Italy and movement between them, for mythologizing various flows between Africa and Italy, and for responding to the American/Hollywood cultural presence in Italy. In some respects, I was interested in describing how certain places became designated as "popular" and how they became "common" as part of Italian cinema's historical relation to a changing conception of the national and the popular in Italy. I tried to underscore that even though one could chart the repetition of certain "common places" (or *topoi*) across film narratives and genres, a particular film's attention to particular generic or mythic places was predicated upon that place's relation to other *topoi* in the "atlas," as it were, of film culture. Cinematic treatments of rural Italy could not be understood except in their relation to treatments of other places. Often a single film would work through the implications of passing from one of these sites to another. Because, however, I wanted to locate the production of a "national–popular" in the complexities of film culture (and, in so doing, to avoid *equating* these films with Fascist initiatives or suggesting that the films simply reflected all-too-familiar events in those years), I may have simplified how the national–popular "gained definition," in a spatial sense, and how it was fraught with tensions and contradictions across multiple sites of cultural production.

I want to engage, therefore, in a brief tactical exercise. Let us take, for instance (and I do not mean to be disingenuous about my choice), two Italian films, *Sole* (Sun) and *Terra madre* (Mother Earth), both directed by Alessandro Blasetti – the first released between 1929 and 1930 and the second in 1933. Both concern urban, male characters whose efforts to "rehabilitate" rural Italy lead them to be seduced by the films' landscapes (the films' characters constituting part of those landscapes for the urban characters). Against or alongside this perspective are the aspirations of the rural folk whose vision of a *cultivated* landscape is mediated by the urban figures.

The latter film reworks the former's narrative. In *Sole*, an engineer's mission to drain the Pontine marshes leads to his becoming embroiled in the struggle of a rural people to reclaim the land. *Terra madre* emphasizes the *return* of a young aristocrat to the land (his family's rural estate). Instead of pursuing this issue of genre or myth formation, however, I want to move laterally, piecing together a surface and network of cultural production through which these films emerged. The rural–urban perspective upon which these films play (and which they attempt to mediate) opens onto a variety of flows that were reshaping the Italian landscape and reconfiguring the nation as a popular territory.

In part, these were flows of people from the provinces to the city, though Italy's population remained relatively more provincial than in other European countries. The flow of people from the country to the cities encouraged two kinds of spatial projects. One involved the emergence of Italian architectural "rationalism" which idealized the utilization of reinforced concrete to produce an infinite repetition of blocks and cells. While this style became most evident in the construction of public administration facilities, and in general as a means of "rehabilitating" urban centers with an eye toward greater regulation, it was also seen as redressing the problem of a "public" or "popular" housing. The most famous example was in Rome in 1931, though Milan's increasing military industrialization over the rest of the 1930s accelerated planned, "popular" housing districts on a more massive scale. The second project had to do with plans for modern cities in rural areas ("rural cities") with the aim of "depopulating" the increasingly crowded cities. The Fascist initiative to drain and develop the Pontine marsh area outside of Rome, which *Sole* references explicitly, lay somewhere in between these two grander projects and flows of people.

Contiguous with Blasetti's two feature-length films and also these land reclamation and urban planning projects were documentary films and newsreels which not only reported on events of national importance in specific places (e.g. an agrarian festival in Forli's central square) but also were required by law in the early 1930s to circulate nationally and to accompany feature-length films. A particularly significant example was a Cines Documentary's treatment in 1933 of the newly erected "rural city," Mussolinia. The film not only works to establish an organic link between the city's construction and the countryside but also notes the place of the city's movie theater as one of its modern features.

What interests me about this film is, first and foremost, its role in locating and mapping the relation among points in a national space. However, in order not to over-emphasize film narrative, it would help to move from these films back to the spatializing implications of their flow. To do this, one could consider the location of film theaters amidst the land reclamation and urban planning projects mentioned above. I am referring to a wide variety of relations: to the urban location of first and second-run cinemas – in urban centers vs. its expanding peripheries; to the practice of outdoor vs. indoor screenings – and the way that outdoor screenings transformed public spaces; to the practice of traveling cinema caravans that again tied the circulation of cinema to a project that was both national and provincial. Connected to this rural cinema initiative were programs to restore ancient outdoor theaters with the effect of transforming a rural landscape into a backdrop for a theater that was "popular" by virtue of its lack of enclosure.

The remapping and reterritorialization of the "national–popular" also occurred through travel and touristic practices that defined and were defined by their relation to the cinematic in the space of everyday life. Early initiatives toward a highway system begun in the 1920s around Milan were taken over by the government in 1930, and streamlining of the railway system was also accelerated (the famous dictum that the Fascists made the trains run on time). Both projects were urban-oriented, emphasizing travel to and among cities while transforming the countryside into scenery for urban travelers and tourists. Amidst the rail and highway programs were the *treni popolari* that carried urban inhabitants into the countryside. Beside the various newsreels and travel documentaries that mapped rural sites, feature-length films mythologized touristic paths as vectors in a national–popular space. The most convenient example is R. Matarazzo's *Treno popolare*, though films such as *Palio* and *Assisi* also have the effect of placing smaller towns on a national map – an instance of film's implication in a growing national practice of tourism.

The spatial politics of flows *in* Italy becomes more complex when they are recognized as part of flows between African colonies and Italy – the relation of flows of film crews and the flow of arms increasingly becoming instrumental in a process of mapping and territorialization. The planning and engineering of new rural or "mining" cities *in* Italy occurred through the same kind of movement of workers that ventured off the peninsula to North Africa to inhabit militarily- and cinematically-defended settlements/outposts. Documentary and feature-length films shot in the African colonies employed these workers, soldiers, and colonial subjects to produce the cinematic "maps" of this territory that were processed in and disseminated from Rome – the new City of national cinema (*Cinecitta*) and the center of Italy's New Empire.

The imbrication of films such as *Sole* and *Terra madre* in a spatial reorganization of the Italian landscape and territory (i.e. as new territorial sites/sights) was, significantly, a process that recast the conditions for a national–popular *history*. While both films were set in present-day Italy, they both perpetuated and revised the conventions of the Italian historical film, arguably the most internationally successful film genre before the First World War. And as Bakhtin has argued about the novel, narrative space is enabled and constrained by a cultural topography – a chronotopic frame of reference – which has been relatively established through preceding texts and which has thus already conditioned the context wherein new narrative spaces (and topographic models) are produced. *Terra madre* not only reworks *Sole*'s distinctions and homologies between the place of the peasantry and the place of the ruling gentry against a rural landscape, but the narrative ambience of these distinctions operates within other residual and emerging chronotopes – "opening," for instance, the conventionalized emphasis upon the "boudoir" and "drawing room" in Italian silent film, or displacing the representations of class distinction through urban settings in contemporaneous films such as *Rotaie* (1929–30) or *Canzone dell'amore* (1930).

To argue that chronotopic reference conditions the context wherein narrative spaces are produced does not adequately explain however how the emergence of cinema, as a gradually institutionalized site, was integral to the realignment and spatial redefinition of an environment – how, in this way, it recast the conditions for a national–popular history. To understand this process, one would need consider how, in "revitalizing" the Italian cinema in the early 1930s, these two films were produced, circulated, and engaged amidst the construction in Italy of new rural cities, urban planning, and expansion into Africa. The design concept of the *casa rurale* (the Modern ideal of the rural habitat) was, through rationalist architecture, brought into a relation with "public" administrative buildings such as the *casa del Fascio* erected in many Italian cities by the early 1930s as a "recovery" of Italy's past for a Modern urban environment.

To see the formation of a national–popular space (or to see the national–popular in spatial terms) points to a kind of Utopianism underpinning a Fascist administration of power. I am using the notion of utopia here (in the archaic sense, an "everyplace" or "no place") to refer to the technological elimination of distance and difference among places. This is certainly a moment when Rome emerges as a center around which a utopian model of nationhood takes shape. The planning and construction of *Cinecitta* ("Cinema city") on the outskirts of Rome after 1934 was a project that had just as much to do with Rome's emerging status as a center of film-making as it did with the planning and construction of other "model" cities throughout Italy and with other objectives of reclaiming the land around Rome, of expanding its political and cultural network.

I also want to suggest, therefore, that the centralization of power was a response to flows that were not entirely cinematic and that were not entirely directed by the State or governmentality. And just as importantly, future research could begin with another – perhaps more provincial – location than Rome in order to consider how it was intersected by various flows (some of which were cinematic) that transformed its internal and external relations, that brought it into a new, perhaps conflicted sense of its relation to locally ritualized spatial models, and that affirmed and complicated the relation among sites and subjects to a Utopian grid or spatial display of national sovereignty and empire. (This is, I think, what Fellini attempts in *Amarcord*.) It is precisely in recognizing the spatially *situated* and spatially *articulated* features of utopian models that one can begin to discuss the conditions upon which the formation of a "national–popular culture" (constituted partly of film practices) emerges historically. It is in recognizing that social relations are organized spatially that one can begin to see film as a historical *site of power*. And it is precisely in recognizing film's exteriority – its relation to a terrain of spatial practices that in turn spatially define meanings, uses, and the place of the "cinematic" – that one can begin to discuss the conditions upon which film produces social relations and thus makes History.

THE LAST PICTURE SHOW IN BOLOGNA
(WHAT REMAINS OF THE CINEMATIC CITY)

In order to avoid a false dichotomizing of History and Space or of Film Aesthetics and Materiality (where history refers to the transformative features of filmic texts and aesthetics, against the immutable features of a material environment), it is important to consider film as a social practice in its multiform relation to particular sites and flows (and social subjects' relation to them) and to consider simultaneously how these sites and flows condition and are conditioned by the *reshaping* of landscape, environment, and territory (and social subjects' relation to *them*). Landscape is thus a concrete surrounding that both limits and diffuses cinematic agency, but it is also constantly worked upon and changing shape, providing different sets of conditions for social agency. Landscape, as that which surrounds the cinematic, makes cinema's potential effects impossible to understand solely in terms of "the cinema," and it is only within this landscape that one can begin to look for the positions from which cinema can be said to have an effectivity. The techniques of projection and observation involving the cinematic apparatus, for instance, may be traced to the *camera obscura*, but the "cinematic" has a historical effectivity only when its dispersal and localization through particular sites contribute to a broad reconstitution of an environment.[20]

Dwelling upon how film depicts and mythologizes landscape or how it brings a spectator into its spatio-temporal diegesis always risks addressing cinema's effectivity purely in terms of form and consciousness, as if they somehow transcend a concrete environment or as if their effects were so uniform as to make their environment (and cinema's place in it) irrelevant. The question of cinema's effectivity should instead be posed this way: how have certain sites been designated and (however tenuously) institutionalized as "cinematic" within an environment wherein the cinematic has (or is recognized as having) a potential to effect certain kinds of change – wherein its relation to other sites matters and makes a difference, both in film subjects' connectedness to and mobility among other sites and in the ideological and material (re-)constitution of social relations as an environment? This is not to ignore that the change may be conducive to a Fascist administration of power, as was the case in Italy during the 1920s and 1930s. Nor, in thinking about power this way, should Italian Fascism or Italian cinema during the 1920s and 1930s have to be conceived as a Utopian model of social–spatial control. This is, however, a more concrete way of thinking about power, and particularly the power of cinema, than in earlier film criticism that understands its project largely in terms of filmic representation and ideology.

If cinema can only be understood through particular sites, then we also need to think about how a cinematic connection to certain places transforms those places, their relations, and social subjects' relation to them (not only their perception and conception of places, but also their access to, investment in, and movement through them). The "cinematic city," understood this way, is a strategic articulation. In part this is a critical strategy interested in what the cinematic foregrounds about the organization of social relations into cities (places)/"cities" (concepts and maps) and in what the urban foregrounds as one environment wherein the cinematic has been organized in particular ways, with particular kinds of effects. But this critical strategy presumes a historical dialectic between cities and cinema, which during the twentieth century strategically shaped both in particular ways – an articulation strategic to the formation and transformation of the cinematic and the urban. And while this articulation is about how the cinematic has been "urbanized" or how the urban became cinematic, it is not about how the cinematic was *essentially* an urban cultural form, or how urban culture was *essentially* cinematic. Instead it sees the cinematic as one formation that has mediated the city (as the material structuring of land) and the "city" (in Lefebvre's terms "the urban" – the social reality, image, concept, myth, spatial model, mapping of the city as place).[21]

Certainly by the 1930s, the dispersal of cinema (and efforts to regulate both its spread and its new relation to an environment) make it necessary to recognize the "cinematic" features of that environment. By the 1930s, the place of cinema had become inextricable from a broad (re-)configuration of urban space – its leisure space, public and private space, civil and governmental space, class and neighborhood space, gendered space. Within urban space (or across the various sites comprising an urban landscape), cinema could be said to have a historical effectivity. But, in turn, the "cinematic city" bore an important set of external relations – to rural life, to regions, to the nation as territory, and to empires and international alliances. In this sense, the cinematic city was bound to a historical network of sites and flows that have since been reconstituted, and in some places disappeared. Where is cinema now? What remains of the "cinematic city"?

I want to address this question by returning to my initial consideration of historiography in cinema studies and the recent Centennial celebrations. Part of what I am reacting against concerns cinema studies' predilection for discussing film form or cinema – *its* subjects and *its* history – as discrete objects and in terms of insulated sites of cultural production. Cinema studies' tendency to do this has, as noted above, been particularly evident in its treatment of "classical cinema" and in discussions of the cinematic apparatus and a revolutionary film aesthetic. In these studies, cinema is never understood overtly as a relation among sites and in terms of the relation of cinematic sites to other sites. Though one could attribute this tendency to a variety of conditions, some of which I have mentioned and many of which lie beyond the scope of this chapter, it has had to do with efforts to locate cinema studies within academic disciplines and at a moment when the place of cinema in social relations was undergoing some significant transformations.

In one respect, efforts to theorize and analyze a more recent history of film production has involved greater recognition of the material dispersal or "fragmentation" of Hollywood cinema. New Hollywood, understood as a mode of production, is seen as a consequence of the erosion of a classical studio system and as a "post-Fordist" realignment of cinema production and distribution through multimedia entertainment industries. Or New Hollywood is seen often as an *aesthetic* loosely related to a post-Fordist restructuring of the industry but directly symptomatic of "postmodernism," understood as an aesthetic or as *the* "cultural logic of late capitalism." Still, emphasis has remained upon film or media relations, assuming either that New Hollywood is only about cinema's relation to a multimedia environment or that cinema is just one of many media constituting an all-encompassing "postmodern" or virtual landscape. Postmodernism (and modernism, for that matter) are often deployed as universalizing terms that generalize the complex specificity of spatial relations and the politics of spatial practice; theories and demonstrations of postmodernism replace the earlier dichotomy of the discursive/non-discursive with a view that everywhere sees only their implosion – an empty landscape or "hyperspace." (And this argument pertains as much to those still occupied with issues of aesthetics and textuality in media studies as it does to a recent vein of "critical geography" that has used such concepts of postmodernism to theorize place and space.) It is one thing to argue that current film or television texts project a "postmodern" landscape, or that there is little difference between the media image and the image-glutted landscape to which it refers, or that the architectonics and redevelopment of urban areas has (as Venturi once argued) come to emphasize display value. It is quite another thing to try to understand the relation between the cinematic and an environment as mutually determining and constitutive (though not in an unconflicted way) in different contexts. What I am outlining is not interested either in discovering a new aesthetic or cultural logic, reproducible everywhere, or in a critical geography that looks anywhere and finds variations of a cultural logic. In fact I am only interested in geography, architecture, and urban/landscape studies to the extent that they provide certain vocabularies and concepts through which one might think about media in social relations – not something that many of the people invested in those fields have readily addressed except in generalized terms.

One of the biggest ironies of the need to celebrate the history of cinema is that the celebrations occur just as realignments between cinema practices and other media practices make the location – the spatial referent – of cinema more of an issue and render the writing of a recent history of cinema – as a discrete object – more problematic. These celebrations affirm that cinema is a disappearing object while attempting to demonstrate that "it" has a history. What the historicism of film studies has ignored is that cinematic practice (including its scholarship) has always been defined through its relations to other sites and that cinematic genealogies are a process of locating and dislocating the cinematic within an environment, of including or excluding some relations rather than others. As I have wanted to suggest, the issue of "film history" (of writing film history, of doing "film studies," of film making History) is a spatial one. Film history can be about little else than *charting the disappearance of an object*. But the task of charting this disappearance involves more than simply recognizing the current alignments and "tie-ins" among media industries; it involves acknowledging the epistemological challenges to cinema studies of pursuing a more radical sense of the "cinematic" as an environmental formation and of the place of its projects in the spatial organization of social relations.

Having acknowledged the current "synergy" among media industries, one might rightly ask to what extent "the cinematic" (or even "the televisual") any longer describes a coherent formation. Are film and television cultural forms whose "centrality" in everyday life and to hegemonic formation has waned? The cinematic, as a dispersion of practices conditioning and conditioned by an environment, was never a discrete formation; nor, for that matter, was/is the televisual. The cinematic has always been connected (in different contexts) with certain sites, its distribution converging with other flows, organizing contexts in that way. Its "coherence" has to be understood that way. The issue of its historical "centrality" in social affairs is a dead end which potentially risks privileging the cinema's determination of social relations or risks taking the ubiquity of a cultural form or a "medium" as a pretext for ignoring the particular location and domain of its effects.

Still, the task of mapping current "cinematic cities" involves identifying certain changes in the cinematic and the city as (overlapping, interdependent) formations. Some of these would include how the "cinematic" has spread into places, such as the home, that it did not occupy in the 1930s, how in this movement domiciles have been reconstituted and realigned, how (largely through "art" and "retro" cinema) the "reclamation" of older urban theaters and public spaces have been part of urban "revitalization," or how it has accompanied the forging of commercial centers (e.g. the shopping-mall multi-plex cinema or, on a grander scale, theme parks). These sites no longer serve only or primarily for movie-watching (e.g. the shopping mall multi-plex or domiciles via television and video playback). And in that sense, cinema may be less recognizable for some social subjects as a distinct site. But together these changes contribute to a spatial redefinition of urban environment as a sphere of social relations and movement.

One consequence of these occurrences is that current cinema production must now map an urban morphology wherein it operates as a residual formation, and from places that it has not traditionally occupied. One way of thinking about a "postmodernist" film style has been as an exercise in superimposing figures from traditional film culture onto a current landscape, or current figures onto a traditional one. To understand the transformation of the cinematic city only in this language, however, ignores how urban environments continue to form around what remains of an earlier cinematic formation, or how the urban is currently an environment organized in some ways by an earlier cinematic formation.

The concept of a "home box office" through television broadcasting attests to lingering conceptions of and distinctions between the home and the movie theater. These conceptions and distinctions are bound to the relation of television production and consumption with that of radio over much of the 1950s. But the transmission of television via cable and satellite, coupled with the use of home video playback, were conditions for and symptoms of a dismantling of a traditional cinematic formation. This process brought cinema into different relations with other institutions and sites even though the residual formation has continued to underpin certain conceptions and distinctions in this emerging formation. The business of film house "restoration" has been key to attempts since the late 1970s in the States to "resettle" commercially unproductive urban centers. Under these conditions, television became a means for some cities to reproduce themselves, historically and geographically. In the competition for tourist trade, TBS and TNT (the Atlanta-based "superstations" owned by Ted Turner) have mythologized contemporary Atlanta, undergoing its own downtown "restoration," through regular national broadcasts of and promotions for *Gone With the Wind*. The example of the Turner-networks' relation to Atlanta also underscores how the advent of television super-stations – through widespread cable and satellite broadcasting – brought different US cities than New York and Los Angeles into a national prominence as centers of "national cultural" production and as touristic centers. The relation of the Disney Channel to Orlando, the Nashville Network to Nashville, and WGN to Chicago have all produced a national visibility for these cities' efforts to transform themselves into sites for travel and tourism. The "televisual" has brought these cities into different relations with themselves, with other cities and with other national and international flows, though all of this has occurred on a territory of flows traditionally and still loosely cinematic (i.e. where Los Angeles and New York have been key coordinates).

The historical overlay between the cinematic and televisual in US cities was thus part of a process whereby the early commercial center of the city, which may have residually functioned in the manner of a medieval European city center, multiplied and spread around the city as part of land development, only to be redeveloped for attracting consumers both from the outlying developments and from outside the city (i.e. tourists). Charting the complexity of how advanced cinematic and emerging televisual formations mediated the transformation of this environment (or how the environment mediated the cinematic and televisual) unfortunately remains beyond the scope of this chapter, and so I conclude instead with an anecdote.

The city of Bologna is a medium-sized city which has retained its medieval organization around a central square, though even that does not make it as worthy of many tourists' maps as are other similarly sized or smaller Italian cities. During the 1950s and 1960s, as part of both the economic prosperity of Italy and the region, the city also enhanced its outlying commercial "fair" – a location which has helped to make Bologna a regional and national center for international trade. In 1996, the fair hosted what was reputedly Italy's first trade show for "new media" technology ("Futureshow"), including "personal" products by companies financially invested in film production and marketing. In the newspapers, coverage of the event repeatedly credited Bologna as Italy's first city for having initiated a project to bring its interested citizens on-line as a means of improving discussion about civic matters. The office for this project is situated alongside other city government offices around the central square, with plans by Umberto Eco to add a Multimedia Arcade as a center for discussion about the creative and civic possibilities of "new media." For the last two summers Bologna's central piazza has also been used as a location where movies are projected free for its citizens and discussed nightly. On the last evening of the series in the summer of 1995 I happened to be in the audience for a screening of *The Last Picture Show*, whose narrative is about the passing of a one-cinema town. The projection of this film in a place which bypassed the drive-in movie theater, but which is becoming a site for promoting internet use and which is engaged by an increasingly well-traveled audience, seems too tempting an allegory to miss for concluding this chapter (and not only because I carry to that film my own recollections of the one-cinema town in Texas where I spent much of my own childhood during the 1950s). Showing a Hollywood film in an Italian piazza is not a recent practice; it occurred during the 1920s and 1930s. But the effectivity of this practice needs to be understood within a convergence between the textual and material features of landscape that has changed.

Modern cities (as Benjamin noted) are palimpsests, comprised of remnants from earlier landscapes, always susceptible to erasure or brought into different relations with emerging structures – social relations redefined spatially as habitat. The cinematic is thus a relation among changing sites where the production of memory for its inhabitants is also an issue of environment. Even though current cities are still environments for, and are still organized through, the cinematic, they are being brought into relations with their own pasts, with their own internal realignments, and with other cities in ways that make the "cinematic city" increasingly a formation whose value to cities lies in the production of the past.

NOTES

1 Henri Lefebvre, "The Specificity of the City," *Writings on Cities*, ed. and trans. Eleonore Kofman and Elizabeth Lebas, Oxford: Blackwell, 1996, p. 102.

2 Jean-Louis Comolli, "Notes Toward a Materialist History of Cinema," reprinted as "Technique and Ideology," *Narrative, Apparatus, Ideology*, ed. Philip Rosen, New York: Columbia University Press, 1986.

3 Though by no means the first, Christian Metz's effort to describe film as a "language" may be the most elaborate and ambitious of such projects. See Metz, *Language and Cinema*, trans. Donna Jean Umiker-Sebeok, The Hague: Mouton, 1974. But my point here has more to do with how semiotics and structuralism, wedded to and understood through psychoanalytic theory, encouraged thinking primarily about the specificity and internal complexity of forms such as literature and film. To the extent that criticism saw its subject as social and/or its project as political (often in its appropriation of certain Marxist concepts), its attention to the internal features of texts encouraged thinking about criticism as intervening in (i.e. deconstructing) the aesthetic by which meaning and consciousness were structured, as making these structures (ideology) visible.

4 Tom Gunning, "The Cinema of Attractions: Early Film, Its Spectator, and the Avant-garde," *Early Cinema: Space, Frame, Narrative*, ed. Thomas Elsaesser and Adam Barker, London: British Film Institute, 1990. Thomas Elsaesser, "Dada/Cinema?", *Dada and Surrealist Film*, ed. Rudolf E. Kuenzli, New York: Willis Locker, 1987.

5 See Miriam Hansen, *Babel and Babylon: Spectatorship in American Silent Film*, Cambridge, MA: Harvard University Press, 1991. See also Giuliana Bruno, this volume, and her *Streetwalking on a Ruined Map: Cultural Theory and the City Films of Elvira Notari*, Princeton, NJ: Princeton University Press, 1993.

6 See, for instance, Colin McCabe's discussion of the "classical realist text" in "Realism and the Cinema: Notes on Some Brechtian Theses," *Screen* 15 (2), 1974 and in "Theory and Film: Principles of Realism and Pleasure," *Narrative, Apparatus, Ideology*, ed. Philip Rosen, New York: Columbia University Press, 1986.

7 The most expansive treatment of this kind has been by David Bordwell, Janet Staiger and Kristen Thompson, *The Classical Hollywood Cinema*, New York: Columbia University Press, 1986.

8 See the first chapter of *Language and Cinema*, op. cit. concerning his distinction between the *cinematic* and the *filmic*. For Metz, the cinematic is what lies "outside" (or temporally, "before" and "after") the film text, but the exteriority of the film text, in this sense, is both 'language system' and 'a socio-technico-economic machine' – a view that makes the cinematic seem self-contained, that generalizes the social, and that abstractly conceptualizes a film's or the cinematic's material environment.

9 See David Bordwell *et al.*, op. cit.

10 See, for instance, Peter Wollen, "Godard and Counter-Cinema: *Vent d'Est*," *Narrative, Apparatus, Ideology*, ed. Philip Rosen, New York: Columbia University Press, 1986.

11 Sigfried Kracauer, *The Mass Ornament: Weimar Essays/Siegfried Kracauer*, trans., ed., and with an introduction by Thomas Levin, Cambridge, Mass.: Harvard University Press, 1995.

12 Roland Barthes, "Semiology and Urbanism," in *The Semiotic Challenge*, trans. Richard Howard, New York: Hill and Wang, 1967. Guy Debord, "Two Accounts of the Dérive," "Introduction to a Critique of Urban Geography," "Unitary Urbanism," and Ivan Chtcheglov, "Formulary for a New Urbanism," all in *The Situationist International Anthology*, ed. and trans. Ken Knabb, Berkeley, Calif.: Bureau of Public Secrets, 1989.

13 Paul Willemen, "The Third Cinema Question" and "The Nation," *Looks and Frictions: Essays in Cultural Studies and Film Theory*, Bloomington, Ind., and London: Indiana University Press and British Film Institute Publications, 1994.

14 This principle, from which I am attempting to distance my own argument, is one that informs Edward Soja's *Postmodern Geographies: The Reassertion of Space in Critical Social Theory*, London: Verso, 1989.

15 See discussions of "narrative space" by, for instance, Stephen Heath, *Questions of Cinema*, Bloomington, Ind.: Indiana University Press, 1981, or David Bordwell *et al.*, *The Classical Hollywood Cinema*, op. cit.

16 While I generally concur with Tony Bennett's rethinking of the relation between the textual and the extra-textual, I find his notion of the "reading formation" as having more to say about the issue of "reading" texts than with texts and readers' relative *place* and mobility within an environment organized and regulated spatially. (See Bennett, "Texts in History: The Determination of Readings and Their Texts," *Poststructuralism and the Question of History*, eds Derek Attridge, Geoff Bennington and Robert Young, Cambridge: Cambridge University Press, 1987.) What I want to propose instead is closer to Foucault's notion that one need rediscover "that outside in which, . . . in their deployed space, enunciative events are distributed" (p. 121) and who describes a "space of exteriority in which a network of distinct sites is deployed" (p. 55). (See Michel Foucault, *Archaeology of Knowledge*, New York: Pantheon, 1972.)

17 James Hay, *Popular Film Culture in Fascist Italy: The Passing of the Rex*, Bloomington, Ind.: Indiana University Press, 1987.

18 Manfredo Tafuri, *The Sphere and the Labyrinth*, Cambridge, Mass.: MIT Press, 1990.

19 Mikhail Bakhtin, *The Dialogic Imagination*, ed. Michael Holquist, Austin: University of Texas Press, 1981.

20 See Jonathan Crary's discussion of the camera obscura, *Techniques of the*

Observer: On Vision and Modernity in the Nineteenth Century, Cambridge, Mass.: MIT Press UP, 1990. Crary's discussion is instructive because it refuses to see the camera obscura as a technological practice and ideal that, after centuries of development, culminated in the cinema. His treatment of the camera obscura and his references to the cinematic, however, have more to do with explaining the formation of the social subject in European Modernity than with considering the camera obscura or the cinema as sites within environments (in relation to other sites) or with how the formation of social relations involved social subjects' access to these sites and movement within these environments.

21 This articulation occurs as the material dispersal and locational designation of the cinematic within a morphology of material, environmental structures designated as a city. It also regards the circulation of certain images and narratives within this morphology that contribute to its imagined shape and to social subjects' investment in it as a place to live and as a space that they navigate in particular ways. One perspective presumes that the city is (in Lefebvre's terms) "an immediate reality, a practico-material and architectural fact" while the other perspective acknowledges "the urban" – "a social reality made up of relations which are to be conceived of, constructed and reconstructed by thought." See Lefebvre, op. cit., p. 103.

11

MAPS, MOVIES, MUSICS AND MEMORY
•

Iain Chambers

All revolutions are rooted in dreams.
GRACE NICHOLLS

Home is no longer a dwelling but an untold story of a life being lived.
JOHN BERGER, *And our Faces, my Heart, Brief as Photos*

'So a musical phrase', I said, 'is a map reference?'
'Music', said Arkady, 'is a memory bank for finding one's way about the world.'
BRUCE CHATWIN, *The Songlines*

Maps, musics and memories: the perpetual translation of space – the space of a language, a sound, an image, a life – into the peculiarities of a place, into the shaft of existence constituted by the passing 'now', inevitably invokes the translation of geography into ontology, of the formal grammars of music and image into the inscription of being in the event of hearing and seeing, of the abstract into the body. I have chosen to consider these questions in the metropolitan context of aural maps, in the recording of sound, in the metaphorical and metamorphosing power of music – giving 'voice to the enigma' (Ernst Bloch) – in the epoch of global technological reproduction. The focus of the essays in this book has concentrated on cinema and the city. I propose to adopt that perspective not as a point of arrival, but as a point of departure for a further set of questions while employing aural, rather than visual, maps to orientate my journey. However, since the testimony of film and cinema clearly remain central to such considerations, let me briefly commence with the question of cinema. As a language, as an economic and cultural institution, a way of picturing and enframing the world, cinema contributes to the making of the visualscapes, soundscapes and culturalscapes in which we move. As such it is also 'a repository of our knowing and our memory'.[1] This perhaps suggests that we should not readily subscribe to the mimetic fallacy and rush in to seek a causal or immediate connection between cinema and society, in which the former is presumed to mirror, reflect and represent the latter, but should instead consider cinema as one of the languages we inhabit, dwell in, and in which

MAPS, MOVIES, MUSICS AND MEMORY   231

we, our histories, cultures and identities, are constituted. To ask what cinema is is to ask what our culture is, who we are, and what we are doing here.

To propose this moment of reflection is to oppose theoretical seizures of the world that seek to reduce language, be it of the cinema or everyday speech, to a transparent medium that is deployed in a second moment, after the subject. The idea that language comes after the subject, rather than constituting something that is already waiting and calling us, is a proposal that does not allow us to travel very far into the entwined enigma of cultural speech, historical being and social becoming. Languages, whether literary, cinematic, musical, or verbal, and even if often dependent upon quite precise techno-cultural systems, are not turned on and off by the flick of a switch. They persist and permeate our world. They ghost our presence and circulate beyond our individual volition. As part and parcel of the ecology of our lives they exist prior to our knowing and thus inform our being and becoming. They are irreducible to a medium or technology. They are part of our understanding.

Where, for example, does the cinema, television or the record end and the social begin? Or, in another lexicon, where does the commercial commence and the aesthetic conclude? The impossibility of defining such boundaries draws us beyond the narrow distinctions that seek to maintain such mediums, such languages and technologies, at a distance, whether critical or social. We cannot withdraw from them, they are always at hand. On the contrary, we are forced to recognise their ontological centrality to who we are, and to what we might desire to become.

Perhaps this lack of distance, resulting in a propensity to be enveloped and made over in these languages so that they represent nothing other than the ambiguous journeying of our being in the world, is most acutely signalled, though rarely considered, in the dominion of sound. It is here where the immediacy of visual regimes and the surveillance of ocular sense is usurped by the infinite relay of song and silence where listening can become as significant as sound. The potential ambiguity, and subsequent freedom, of the image and film is not denied, but institutional and economic pressures, even the ideological premises of factuality embodied in common-sensical understandings of the visual, rarely permit that promise to be. It is in the wider, dispersive context of sound, I wish to argue, that such a promise is more fully maintained.

Although continually embedded in appearances, in the visual economy, the body continually exits from this daily frame through the migrations of sound. Memory clings to the former while following the itineraries of the latter. In the unique instance of the performance, sounds chafe against the constrictions of ocular hegemony – this look, that style, those bodies, that moment – constantly threatening to break bounds and travel without regard for address or direction. The visual fetishisation of the object, its fading aura maintained in deteriorating celluloid (although now destined for the glow of an after-life in digital archives) is consistently carried further afield by sound – notes, shouts, respiration, the rap, the silence: the space that offers hospitality to the future. The scopic drive, the 'subject' that constitutes and projects its 'object' and seeks to render all transparent, scientific, clinically apprehensible and mastered – herein lies the uncanny proximity of medical and media discourses – is invariably intent on grasping being and time, turning life into an exemplary instance, an abstraction, an ever-ready 'standing reserve' of meaning.[2] The image, for all its potential ambiguity, tends towards the potential consolation of a semantic full stop. It is sound that ultimately disturbs ocular regimes and returns images to the pleasure of surfaces, to the liberty (and limits) of the making, masking and masqueing of representations. This is to contest the triumph of the image over the act, of the disembodied form over the corporeal flux, of the metaphysical signature over the unruly event, of the sign over the sound.[3] The immediate factuality of the image – the photograph, the cinematic sequence, the digital scan – apparently embodies and exhausts the presence of being. In freezing time, while paradoxically recording and replaying it, the image is linked to death where everything can seemingly be exchanged as relics of being in the flat circulation of signs and their ghostly after-life. Returned to the realm of the senses, however, the image might be better considered to encapsulate the mechanisation of representation: the technological visualisation of writing. By exceeding such a writing, imagining and framing of the world, the visual conclusion is relocated and the image is forced to reveal its logocentric impulses as a power and a limit, as a promise and a threat, as an extension and a closure. Between this and what continues to exist outside the frame lies a path along which the *poesis* of sound maintains the promise of the irrepressible.

A voice in the dark, a saxophone cadence on the street, saliva on the tongue, breath drawn between words, the suspension of silence: all this is music, and all this mutates immediate motives and punctuates them with memory. To ask the meaning of music, the significance of sound, is perhaps to seek to distil from the depths of our senses the ungraspable beingness of being. George Steiner writes: 'In music, being and meaning are inextricable. They deny paraphrase. But they are, and our experience of this "essentiality" is as certain as any in human awareness.'[4] Weaving together beat and being, sound and memory, the desire and deferral of musical meaning hums an internal counter-point in the pre-linguistic state of language, in the indeterminable semiosis of our bodies.[5]

Music permits us to travel: forward in fantasy, backward in time, sideways in speculation. Above all, music draws us into the passages of memory and its 'sudden disjunction of the present'.[6] Here time overflows the containment of our concepts. The time of memory is a reversible time that permits the return, revisiting and revision of other times. The coeval presence of this reversible time and the irreversible nature of our bodies opens up a rent in our experience, in our lives. Linear, irrevocable time is interrupted by the interval and intrusion of transversal times: the genealogy of the symptom, of recall, of the recording and the reordering of the past, and the perpetual desire to return to the record. The mutable place of memory erupts in the space of our histories as a set of fragments suspended in time, as a loss that is experienced as essential to our comprehension of the present. Memory is sustained and held in custody (both captured and protected) by the fragile chains of language, by the rhythms and respiration of the body that constitutes the ambiguous aperture and agenda of our identity. And music, as song, dance and rhythm, as musical maps and song lines, forms a contrapuntal memory that sounds out circumstances in the creation of a mobile individuation and community.

A gypsy girl dances alone beneath a desert palm. Her brightly coloured dress and jewellery flash in the evening shadows. The scene is both hypnotic and emblematic. We find ourselves on the threshold of an elsewhere that opens our world to the disturbing presence of something that we recognise but which flees our desire for comprehension. Here there is something that evades the closure of our understanding. Everything is clear – the figure dancing in the desert, the sounds that accompany her movement – but something remains opaque, hidden, out of sight, mute. This is the opening scene of the film-journey of *Latcho Drom* (1993) by the Algerian director Tony Gatlif. The journey commences in the north west area of the Indian subcontinent and goes on to traverse an archipelago of historical memories that are revealed in the sounds that emerge along the way: Pakistan, Egypt, Auschwitz, La Camargue, Andalusia. This history of a people without a home illuminates in a flash Heidegger's proposition that language is the house of being.[7] This lateral history is maintained and nurtured in the interminable journey of a changing song that passes from horizon to horizon transforming the passage in music, the earth in history, a diasporic space and identity into places and the house of language. 'Vertical' history, dependent upon the ground we find beneath our feet in the ancestral soil, is here displaced by the persistent travelling of a language – incorporated in sound, in song and dance – seeking an accommodation in the world. In this nomadology of sound, this musical inscription of the earth, I become aware of a state in which music and the world meet and combine in the body, rendering explicit the alterity that makes each of us a subject.

Identity itself is a shifting, combinatory figure, a musical phrase in the score of being. The sound begins, we are born, but once under way it does not conclude until mortality imposes a coda. Identity is an ever present, ever unfolding, bass line; a rhizomatic figure, a fugue drawn from the languages that transport and sustain us, a solo and improvisation on the energies that unfold and devolve in the world (rather than an isolated note that withdraws and redraws the world into the *subiectum* of the self).

In the recent history of the West it is the insistence of Freud that returns language – whether as irreducible ontology (Heidegger) or symbolic constraint (Lacan) – to the body, to the sounds of being.[8] To consider music as memory is to grasp the vital and physical nature of repetition; of how, according to Freud, remembering (*Erinnerung*) is linked to repeating (*Wiederholen*). In 1914 the father of psychoanalysis wrote a brief essay entitled 'Remembering, Repeating and Working-Through'.[9] He notes the importance of repetition in discharging symptoms 'along the paths of conscious activity'.[10] Further, he underlined that repetition can provide both an access to memory and a mode for resisting, refusing and repressing it. Music, as a language of repetition, continually proposes this play between recalling and resisting the past. In the return of sound music fills in the intervals in memory, provoking a temporary overcoming of the resistance to its presence and the body that incorporates it. In the instance of repetition perhaps it is not so much the case of remembering what has been forgotten as of exposing the act of forgetting itself? Oblivion is forgotten, but the language of repetition simultaneously takes it in hand and transforms it. This continual song, 'not as an event of the past, but as a present-day force', provides consistent custody for the presence–absence of the memory of being – revealing in the event of sound that simultaneous disclosure and concealment that is the fundamental role of art according to the thinker of being.[11]

Memory, around which so much of the sense of our selves revolves and returns, is the skin stretched over the world across which desire, emotions and expressions flow. Memory evokes the eroticisation of the past. But memory does not exist as an autonomous realm; it is sustained and guarded by language, in the record of images, words and sounds. Contesting the apartheid of memory, and the agents of oblivion seeking to consign the past to the conspiracy of silence, music sustains an ethical resonance that permits us not so much to understand and interpret the past as to recover fragments of its dispersed body. Beyond the rigid monologue of reading, cataloguing and interpreting antecedents, music establishes a potential site of simultaneous interpellation and response that directs us elsewhere. Music allows us temporarily to invert the course of time and to consider history as a reversible testimony – hence never assured, as Benjamin points out – that bears witness to the past, to the ruined redemption of humanity, to a sorely tested but eternal faith in our habitat. Perhaps music, by drawing us through the gaps in time and meaning, allows us temporarily to exit from the narratives that frame us in order to re-negotiate our 'home' in them.

Our access to memory is through language, through the traces inscribed on the page, our bodies, and in the auditorium in which we speak and listen. Not only do we recall our past in music, but the very techniques that permit us to return there, recordings, are a form of inscription, of writing. In *The Aesthetics of Recorded Sound*, the contemporary Japanese critic Shuhei Hosokawa writes: 'It is not by chance that incipient devices such as the phono-graph, gramo-phone, and grammo-phone were all given names derived by combining "sound" and "writing".'[12]

Displayed in the instantaneous languages of photography, film, record and digital images our memories are increasingly rendered proximate. There they are captured, amplified and dispersed. Reproductive technologies and techniques permit an 'eternal return'. But there also remains a profounder strain that no technology or technique will ever be capable of fully translating. In the desire for time, for life, our memories reach out to protect us from oblivion, and in the shifting but repetitive modalities of our subscription to sound that clear, but indecipherable, desire attains the maximum of ubiquity. Music serves as a multi-dimensional map; it is simultaneously connected to fashion (repetition of the new) and to memory (moments lost in time). It permits us to maintain a fragile bridge between consciousness and oblivion. It introduces the history of the event into the fluctuating, atemporal regime of memory by permitting us to mark time and recall it, admitting the past to the present, and allowing us to trace in its echo other dreams, further futures.

So music, although initially the expression of a historical and cultural instance, once released into the world travels interminably; it has no single location, it continues to continue 'without why' (Master Eckhart). It is everywhere and nowhere: the hole in time, the slash in space, the noon-time of experience.

But although ubiquitous, sounds are always transcribed into the particular poetics of a place. As language, as writing, memory, music and the murmur of being always involve an act of translation. In the transfer, in the re-membering invoked in the transit of translation, the intention of re-presenting something that previously existed is interrupted. It is overtaken in a process – the historical work, the imaginary work, the dream work – that transforms as it transports something from one place to another and in the movement displaces it.[13] The desired mimesis between reality and representation, between past and present, is deviated by the radical historicity of the situation that exists in the perennial gap between the excess of sense and the limits of each instance of translation, memory, writing, narrative and recognition. Translation reveals the disruption in the very foundations of translation. It cannot speak in terms of universal meaning and the transparency of truth, but rather in the accents of the cultural configurations and social interstices in which language, representation and reality are destined to be transcribed, transformed, and made over again and again as it struggles towards an accommodation.[14]

There exists no simple or direct recovery of how things 'really were', only of how things come to remembered and translated, not what happened, but what is happening. So, everything is both remembered and repressed, every testimony is flawed, every recall destined for another return. Memory is also the art of forgetting, of occluding a loss, negating a lack, cancelling the failure of language, the incomplete destiny of the intention. Memory is thus not an origin, a prescription, a destiny, but a resource, a table of writing . . . in which the poignancy and pain of the past is overwritten in the psychic release of the present.

Memory, whether personal and psychoanalytical, or collective and historical, dwells in an ambiguous landscape. For Walter Benjamin memory and history were interchangeable.[15] Does that mean that memory is only a collective historical configuration that excludes any sense of the obfuscating slippage and repression that betray the individual's unconscious? Or could it also suggest the idea of a historical unconscious where the explicit, rational account is forever shadowed, doubled and displaced by another story, another scene? And does that second path not also permit the possibility of contemplating an individual configuration of the past in the suggestive, but diverse, manner of Proust or Benjamin's own Berlin childhood? It is history not as science, but as memory; not as the law, but as language.

Memory is neither fixed nor eternal. It mutates. As a cultural constellation it permits an oblique glance that scans historical time diversely, proposing a redemption in which cultural speech and historical action are re-written through the introduction of the unstable power of metaphor and modes of language that disrupt as they disseminate discourse. This, for example, is what provided Benjamin with the method of turning history against its own provenance. (Here also lies his affinity with the disenchanted baroque poetics of the shipwreck of words crashing against the unnameable, the swirl of notes drifting towards infinity.) If history and memory share the same structure, then history merges into a setting: not into the prescriptive regularities of a linear structure, but into the discontinuous inscriptions of the scene. So, Benjamin's reading of images necessarily involves a breaking of images, a critical awakening through which he seeks to recover, collect, and prise them loose from an uncontested continuum. This, too, is the case with memory: it is necessary to break into it, to interrupt its consolations, and 'to discover what the present might have been by a re-cognition of the past.'[16] In the secret agreement between past and present the body of history, the history of the body . . . history as body, breaks into the past in order to re-configure and em-body the present in an ontological interruption.

Both romanticism and Marxism ground themselves in the a priori of a non-alienated society: a reality that has been obfuscated and obscured. Both criticise the present for a lack, an absence; for its failure to restore us to a lost, organic unity. But what if alienation, like contradiction, is not peculiar to capitalism? What if alienation never experiences *Aufhebung* and is a terrestrial constraint irreducible to the social relations of a particular mode of production and therefore continually defies the 'progress' projected in all teleologies? In other words, what if contradictions are not teleological but rather ontological, holding us prisoner in the time of our being? To recognise this condition, while struggling against it, means less to push forward and more to move sideways by refusing to perform the prescriptive by reworking, re-routeing it differently. It is to cast oneself not forwards towards utopia (itself the humanist child of a eurocentric 'progress' set upon discovering new worlds), but sideways into atopia: another place, a diverse way of inhabiting the world. The utopic is usurped by the heterotopic, by the proliferation of space into different places, languages, rhythms, events . . .

The inferable essence of music perhaps best helps us to elude the rational insistence of 'progress' in our traversal journeys sideways into the expansion of the present, into the architecture of sound employed in building imaginary homelands, constructing temporary habitats and seeking accommodation in the world. Music, as dance, as carnival, as rave, as ecstasy, draws us out of our assigned social space into a sublime place. It permits a transitory exit from the dance floor to the trance floor, thereby shortening the passage and facilitating the access to an elsewhere that previously required much more time, money and existential conviction, whether that had meant becoming a member of a spectacular subculture, an aesthete, or else completely 'dropping out' into an alternative life-style.

It is memory that protects us from the past which would otherwise flood in and swamp the present. As a construction, a shelter we build, a tale we tell, memory serves both to brake and to bestow shape on the inchoate that would otherwise overwhelm us. In a self-portrait entitled 'The Shadows' written in 1925 for Martin

Heidegger, Hannah Arendt writes: 'All sorrows can be borne if you put them into a story or tell a story about them.'[17] Memory is not merely formed, it is framed and, in a subtle sense, faked. It provides a site which seemingly offers access to the past but which can never escape the interrogation and shape of the present. Memory is a mechanism that selects. In our recall we re-member certain things while forgetting, negating and denying other things. Our memories are as authentic (or inauthentic) as everything else we do. All memories, all writings, all histories, are fallible and fragmented. Each is a version of what is irremediably lost to it. It is always a re-membering, a temporary putting together of disparate elements of a body that can never be made whole, that is always provisional, illuminated by the light of oblivion. Memory is not a stable monument to the passage of time, but rather an unstable configuration which defies time and which can be attained by different paths to reveal diverse stories. Thus memory is always contested. It can therefore also be the site of amnesia and cancellation. To recognise the past as a scriptable economy, a writing pad or palimpsest, is also to recognise its suscep-tibility to diverse interpretations subject to the powers that seek to authorise the past and, through it, the present. The return and representation of the past is always susceptible to the 'assassins of memory', and thus 'obligates us to become sensitive to the *fact* of the forgotten.'[18]

Music as a language (like all language) maintains this tension through its communal use and individuation. Its ready accessibility compared to other, more formally institutionalised, languages such as literature, historiography and the visual arts, permits a ubiquitous and unexpected punctuation of the scripts we are expected to follow. Music, in its anonymous consumption and innumerable moments of articulation – from the desert ceremony and forest clearing to the bar, street corner, subway exit, and modern consecration in the recording studio – perhaps provides an altogether more extensive and irrepressible configuration of a language that sings time and being while recording memory. If music provides a home for nostalgia, it also offers a point of return for what becomes a new point of departure.

As a construction in and across time, memory is not merely a transparent envelope that contains (and constrains) what we think we know and are able to recall. If it casts our attention towards the past, it also propels us down shadowy slopes into the realm of the unconscious. In this sliding continuum between past and present, where each contributes to the constitution of the other, how can we know what we think we know? What we can, and do, hold onto are the languages that permit us to consider such questions. Such languages need neither be verbal nor explicit. They exist prior to our inhabiting them, and are irreducible to what we say and do in them. Memory is temporarily rendered whole as it replies to the present.

To cite the past is to re-site the present and reveal in it the instance of the contin-gent paths that lead us back while taking us forwards: memories . . . of a Berlin childhood; of a zoological garden; of a city where 'to lose oneself, as one loses oneself in a forest, is an art to be acquired'; or of a Belfast adolescence 'where you could feel the silence at half past eleven on long summer nights as the wireless played Radio Luxembourg, and the voices whispered across Beechie River'.[19] In December 1993, at the Masonic Auditorium in San Francisco, Van Morrison sings 'In The Garden'. This gently segues into an earlier song – 'Real, Real Gone' – which, in turn, recalls Sam Cooke's 'You Send Me':

> Sam Cooke was on the radio
> and the night was filled with space
> and your fingertips touched my face
> saying darling you send me
> you send me
> you send me . . .
>
> No guru, no method, no teacher
> You and I and nature
> And the Father and the Son and the Holy Ghost
> In the garden
> misty wet with rain . . .
>
> . . . did you get healed tonight?
> . . . did you get healed tonight?

Another time, at the end of a dark alley in the Quartieri Spagnoli in Naples in an eruption of light stands the Galleria Toledo Theatre. I have come tonight to see and hear the Algerian dancer El Hadi Cheriffa accompanied by the voice and percussion of Moussa Belkacemi. The sounds, the body, the poetics of a visual and aural grace harvest traditional music and forms in a volatile mix (Bedouin, Berber, Tuareg) whose distinctions and composition are equally part of the modern Maghreb. The question of their composition, and the composition of such a question, shadows the dance, echoes in the song . . .

> The voice that sings
> like the hand that writes
> is the body in language
> responding
> to the call
> and
> care
> of being.

NOTES

1 Tony Fry, 'Introduction', in T. Fry (ed.) *R/U/A TV? Heidegger and the Televisual*, Sydney, Power Publications, 1993, p.12.

2 Martin Heidegger, 'The Question Concerning Technology', in M. Heidegger, *The Question Concerning Technology and Other Essays*, New York, Harper and Row, 1977. This is not to negate that the language of cinema cannot provide new beginnings in contexts where images also become historical witnesses and ethnographic signatures for something and someone else, establishing a difference within the increasing global belief in the visibility of truth. See Rey Chow, *Primitive Passions, Visuality, Sexuality, Ethnography, and Contemporary Chinese Cinema*, New York, Columbia University Press, 1995.

3 For a courageous and thoughtful discussion of the political implications and blindness of ocular hegemony in the context of contemporary rap music, see Paul Gilroy, '"After the Love Has Gone": Biopolitics and ethno-poetics in the black public sphere,' *Third Text* 28/29, 1994.

4 George Steiner, *Heidegger*, London, Fontana, 1992, p. 44.

5 Here the reference is clearly to Julia Kristeva's distinction between the semiotic and the linguistic or symbolic state of language; see Julia Kristeva, *The Revolution in Poetic Language*, New York, Columbia University Press, 1984.

6 Homi K. Bhabha, *The Location of Culture*, London and New York, Routledge, 1995, p. 217.

7 Martin Heidegger, 'Letter on Humanism', in M. Heidegger, *Basic Writings*, ed. D. F. Krell, London and New York, Harper and Row, 1977.

8 The theme of the return of Freud against Lacan is borrowed from a talk given by Julia Kristeva at the Institut Français, Naples, 17 January 1995.

9 Sigmund Freud, 'Remembering, Repeating and Working-Through', in *The Standard Edition of the Complete Works*, general editor James Strachey, vol. 12, London, The Hogarth Press, 1962.

10 Ibid., p. 147.

11 The quote is from Freud, ibid., p. 151; on the 'disclosure' and 'concealment' of art, see Martin Heidegger, 'The Origin of the Work of Art', in M. Heidegger, *Basic Writings*, ed. D. F. Krell, London and New York, Harper and Row, 1977.

12 Shuhei Hosokawa, *The Aesthetics of Recorded Sound*, Tokyo, Keisō Shobō, 1990.

13 For an important discussion of this issue in the vital context of contemporary cultural translation, see Rey Chow, *Primitive Passions, Visuality, Sexuality, Ethnography, and Contemporary Chinese Cinema*, New York, Columbia University Press, pp. 182–95.

14 'even the greatest translation is destined to become part of the growth of its own language and eventually to be absorbed by its renewal', Walter Benjamin, 'The Task of the Translator', in Walter Benjamin,

Illuminations, trans. Harry Zohn, London, Collins/Fontana, 1973, p. 73.

15 Much of this paragraph draws upon a stimulating talk on Walter Benjamin given by Sigrid Weigel in the Spring of 1994 at the University of California at Santa Cruz.

16 Michael Hays, 'Foreword' to Peter Szondi, *On Textual Understanding and Other Essays*, Minneapolis, University of Minnesota Press, 1986, p. x.

17 Hannah Arendt, 'The Shadows', quoted in Elisabeth Young-Bruehl, *Hannah Arendt. For Love of the World*, New Haven and London, Yale University Press, 1982, p. 50.

18 The first quote is from Yosef Hayim Yerushalmi, quoted in Paolo Rossi, *Il passato, la memoria, l'oblio*, Bologna, il Mulino, 1991, p. 28; while the second is from David B. Clarke, Marcus A. Doel and Francis X. McDonough, 'Holocaust topologies, singularity, politics, space', *Political Geography*, 15(6/7), pp. 457–89.

19 The citations are from, respectively, Walter Benjamin's 'A Berlin childhood around 1900' ('Infanzia berlinese intorno al millenovecento' in Walter Benjamin, *Immagini di città*, Turin, Einaudi, 1971, p. 76); and Van Morrison's 'On Hyndford Street', *Hymns To The Silence*, Polygram, 1991.

INDEX

●

Page numbers in *italics* refer to plates; those followed by n indicate notes.